# INDIAN
# TALES AND LEGENDS

7/08
14

# Indian
# Tales and Legends

*Retold by*
J. E. B. GRAY

*Illustrated by*
JOAN KIDDELL-MONROE

OXFORD UNIVERSITY PRESS
OXFORD   NEW YORK   TORONTO

Oxford University Press, Walton Street, Oxford OX2 6DP

Oxford New York Toronto
Delhi Bombay Calcutta Madras Karachi
Kuala Lumpur Singapore Hong Kong Tokyo
Nairobi Dar es Salaam Cape Town
Melbourne Auckland Madrid

and associated companies in
Berlin Ibadan

Oxford is a trade mark of Oxford University Press

First published 1961
Reprinted 1965, 1971, 1979
First published in paperback 1989
Reprinted 1990, 1992, 1993

British Library Cataloguing in Publication Data
Gray, J. E. B.
Indian tales and legends.
1. Indian tales and legends – Anthologies
I. Title II. Kiddell-Monroe, Joan
398.2'0954
ISBN 0-19-274138-1

Printed in England by Clays Ltd, St Ives plc

# AUTHOR'S NOTE

In selecting the stories for this book, the sources have been Sanskrit and Pali, the classical languages of the Brahmans and the Buddhists. Many of these tales were current centuries before the Christian era and were given a sophisticated form by the story-tellers of classical times. Many, too, survive in various forms among the folk stories of modern India. Fundamental in nearly all are the concepts of transmigration (rebirth on earth according to one's deeds in a former existence) and of caste (the rigid barriers relating to marriage and social intercourse set up within Hindu society).

The first five stories are taken from the Pali *Jātakas*, the Buddhist birth-stories, where the former existences of the Buddha are related. In *King Great Virtue*, the king himself is the Bodhisattva, the Buddha-to-be; in *Prince Wicked*, he is the hermit; he is also Suppāraka, and the Master of the monastery in the next two stories. In *The Earthquake* and *The Hawk's Friends*, he appears as a noble lion. Animal fables have always been popular in India and are the source for many of the tales current in the West. *The Price of Greed*, *The Indigo Jackal*, *The Fate of the Vulture*, *Curd Ears* and *The Hare in the Moon* have all been taken from the *Hitopadesha*, a collection of fables, mostly animal, designed to inculcate wisdom and polity. *The Brahman and the Goat* and *The Foolish Brahman* come from the same source. The remaining tales, apart from the two Epics, are taken from the Kathāsaritsāgara (The-Ocean-of-Streams-of-Stories) of Somadeva. All the above are fairly literal translations from the original, with some omissions in the interests of readers, but *Nala and Damayantī* (an episode from the great epic, the *Mahābhārata*) and the *Rāmāyana*, are adaptations.

v

# PRONUNCIATION

In Indian names each syllable should be pronounced with fairly equal stress. Approximate equivalents for the vowel sounds are:

> *a* like the *u* in c*u*t
> *ā* like the *a* in f*a*ther
> *i* like the *i* in p*i*n
> *ī* like the *ee* in f*ee*t
> *u* like the *oo* in s*oo*t
> *ū* like the *oo* in l*oo*t
>
> *e* and *o* are always pure long vowels:
> *e*  like the *a* in h*a*te
> *o*  like the *o* in r*o*le
>
> *ai* like the *i* in h*i*de
> *au* like the *ow* in h*ow*

The consonants, which have been simplified, may be pronounced as in English, but with one important exception: although *ch* and *sh* are as in English, the other combinations with *h*, e.g. *gh*, *dh*, etc. are pronounced as *g*, *d*, etc. followed by a puff of breath, somewhat like the *gh* in lo*gh*ouse, the *dh* in ma*dh*ouse, etc. Note that *th* is never pronounced as in English, but as in ho*th*ouse.

# CONTENTS

## King Great Virtue

IN time long ago there was a young prince who was the son of
the king of Benares. While he was still a tiny infant he smiled
so gently and radiated such happiness that when the day
arrived to give him a name, his parents called him Prince
Virtue. By the age of sixteen or so he had mastered all branches
of knowledge and, some time after, upon his father's death, he
became the just and righteous ruler of Benares with the title of
King Great Virtue.

He set up six great halls, one at each of the four gates of the
city, one in the centre and one right beside the entrance of his
palace, and in these halls he ordered alms to be given to way-
farers and to the poor. He himself virtuously followed his religion, observing fast-days, and, ruling his country righteously,
he kept his subjects happy, as might a father his infant son,
dandling him upon his knee.

On one occasion, one of the ministers grievously offended a
member of the king's family, and the story of it soon became
known. When the ministers brought the matter to the notice of
the king, he looked into it, and, when his investigation was
complete, he had the minister summoned.

'You foolish fellow,' he said, 'you have done a very wrong

thing. You are not to stay any longer in my kingdom, so take your wife and children and whatsoever wealth you possess and go elsewhere.'

Banished from the kingdom of Benares, the ex-minister came to the king of Kosala and, entering into his service, gained his confidence in a while. One day he said to the king of Kosala:

'Your majesty, the kingdom of Benares is like a honeycomb free from flies; the king is very weak and even a small army would be able to seize his kingdom.'

When the king heard what he had to say he thought to himself that, as the kingdom of Benares was of no mean size, anyone suggesting that it could be taken by a small force was most probably in the pay of an enemy, and so he said straight out:

'You have been bribed, it seems.'

'No, your majesty, I have not been bribed. It is the truth I speak, and if you do not believe me, send some men to destroy a border village. If they are caught and brought to the king of Benares, he will give them gifts and send them away.'

The king had to admit to himself that the fellow spoke very confidently, and so, with a view to testing the theory, he sent some men to sack a border village. They were captured and dispatched to the capital. When the king saw them he asked:

'My good fellows, why did you plunder my village?'

'Because we had no means of livelihood,' was the reply.

'Then why did you not come to me?' the king said and, giving them gifts, he sent them on their way with the warning, 'Never do a thing like this again!'

They returned to the king of Kosala and related to him what had happened. The king, still not daring to invade on the strength of this news, sent a further company to make a raid on a region in the centre of the kingdom. These ruffians were also presented with gifts by the king of Benares and were sent away. Even with this evidence the king of Kosala did not dare to invade and therefore sent some men who should plunder in the streets of the capital itself. Once again the ruffians received gifts and were dismissed in the same way. Then the king of Kosala realized what a very good man the king of Benares was and, deciding forthwith to seize his kingdom, set out with an army.

At that time King Virtue had at his command a thousand dauntless warriors. Unflinching before the charge of a maddened elephant, they would have been ready even to conquer the whole of India for the king of Benares. When they heard that the king of Kosala was marching to war, they approached their king, saying:

'The king of Kosala intends to seize the kingdom of Benares. Let us attack him as he sets foot on the border and capture him.'

'My good men, I will not be the cause of hardship to others. Let those who seek kingdoms take them; you are not to go.'

In this way the king restrained them. Soon the king of Kosala had crossed the border and was entering the centre of the kingdom. This time his ministers approached the king of Benares and reported the development to him. Once again the king restrained them from action, but by now the king of Kosala had halted outside the city. He sent in an ultimatum to King Virtue, requiring him either to surrender his kingdom or to do battle. When the king received it, he sent a message in return, saying:

'There is no question of fighting; you may take the kingdom.'

Once more the ministers approached the king and spoke to him.

'Your majesty, we would not allow the king of Kosala to enter the city, but rather we would check him outside the city and capture him.'

As before, the king curbed them and, commanding that the city gates be opened wide, seated himself upon his throne in the great court with his ministers about him. The king of Kosala entered Benares with a huge army. Not seeing a single hostile warrior he went straight to the gate of the palace, and, passing through the open doorways, he mounted to the vast and ornamented court where King Virtue was seated with his thousand ministers.

'Go bind the king and his ministers firmly with their hands behind their backs!' the king of Kosala commanded. 'Take them to the cemetery! Dig holes there and bury them up to the neck! Fill in the earth around them so that they cannot raise a hand! The jackals will come in the night-time and will do what is necessary.'

The usurper's soldiers carried out his orders and took away the king and his ministers with their arms tightly bound behind their backs. Even then King Virtue showed no anger against the usurper, and as the ministers were led away bound, not a single one broke his allegiance to the king, so perfect was the discipline of his followers.

When their enemies reached the cemetery with King Virtue and his ministers, they placed them in holes which they had dug, the king in the middle and his ministers on either side. Then they filled in the earth up to their necks, stamped it firm and went away. The king exhorted his ministers, saying:

'Feel no anger towards this usurper, but rather be kindly disposed towards him.'

Then at midnight the jackals came, anxious to eat human flesh. When the king and his ministers saw them, with one accord they raised a great shout. The jackals ran away in fear. But once they realized that no one was chasing them, they turned round and came back. Again the men raised a shout. In this way three times the animals ran off and, seeing they were not chased, returned. As they became aware that the men were doomed, they plucked up courage and this time did not run away despite the clamour. The leader approached the king while the rest made for the others.

As soon as he was aware of the animal approaching, the king, who was highly resourceful, threw back his head as if offering his throat to be bitten; then he himself bit at the beast and locked it, as in a vice, with his jaws. The jackal, finding itself firmly held and dragged down by the elephant-like strength of the king's jaws, was unable to free its neck and, being scared to the point of death, uttered a tremendous howl. When the other jackals heard that frightened howl they knew that their leader had been trapped by a human and, no longer trying to investigate the ministers, fled in a body, frightened out of their wits.

Meanwhile the jackal which was trapped tightly in the king's jaws twisted and turned to and fro, and the earth was gradually loosened as the animal scraped away for dear life, with all four paws, at the ground around the king. As soon as the king felt the earth loosening he let the jackal go, and, jerking with all his enormous strength from side to side, freed both hands. Then, holding on to the edge of the hole, he heaved himself out, like a thunder-cloud rolling up before the wind. With words of encouragement to his ministers, he removed the earth from around them, releasing them all, and so he stood, in the cemetery, surrounded by them all.

Now it happened that some men had previously brought a corpse to the cemetery and by chance had abandoned it on the line dividing the domains of two demons. The demons were uncertain how to share the corpse and said:

'As we are not able to divide this body, let us go to King Virtue; he is just and he will show us how.'

With these words they dragged the corpse by a foot up to where the king was and said:

'Your majesty, please divide this corpse between us.'

'O demons,' the king replied, 'I will indeed divide this for you, but I am dirty and would like to bathe first.'

The demons, by their magic power, brought the scented water which had been put by for the usurper and gave it to the king so that he could wash himself. When he had washed, they brought him the robes which had been laid out for the usurper; when he had dressed, they brought a casket full of four kinds of scent; when he had put on the scent, they brought all kinds of flowers laid upon jewelled fans in a golden casket, and when he had adorned himself with the flowers, they asked him:

'What else may we do?'

The king indicated that he was hungry. They went and

fetched the expensively flavoured food which had been prepared for the usurper, and the king, now that he was dressed and scented after bathing, ate the food that was rich with many spices. The demons also brought the usurper's scented drinking water in a golden bowl and a golden drinking cup as well. When the king had drunk, and washed his mouth and hands, they brought him the customary delicacy which follows a meal, *betel* nut laid in a leaf and dressed with five most fragrant essences. When he had chewed the *betel* he stood up, and they asked again:

'What further may we do?'

'You must bring the ceremonial sword which has been placed at the head of the usurper's bed,' the king replied, and straightway they brought it. The king took the sword and, propping the corpse upright, brought the blade down on the middle of the skull and clove the body in two halves. Whereupon he gave each demon his share, cleaned his sword and girded it on.

The demons in the meanwhile had eaten human flesh to their satisfaction and, being much pleased, asked once again:

'O great king, what else may we do for you?'

'Transport me by your magic power to the usurper's bedroom and dispatch each one of my ministers to his own home!' the king demanded. They agreed immediately, saying, 'Certainly, your majesty,' and performed the request forthwith.

At that hour the usurper was lying fast asleep in bed in the royal bedroom. As he lay there sunk in sleep, the king struck him with the flat of the sword on his belly. He woke up terrified and, recognizing King Virtue by the light of a lamp, he exclaimed:

'O king, the palace is well guarded by night; the doors are barred; the surrounds are beset with sentinels, so how indeed did you come to my bedside, girt with the sword and fully arrayed?'

The king related then at length the manner of his arrival. When the usurper heard him he was much troubled and said:

'O king, though I am a man also, I did not know the virtue

you possessed, and yet these coarse and cruel demons who feast on human flesh recognized your worth. Never again will I, O great mortal, injure such a virtuous person as yourself.'

As he said so, he touched the sword and swore an oath and asked the king's pardon. Thereupon he placed the king in the royal bed and himself slept on a small bench.

When the night had passed and the sun had arisen, the usurper summoned his army by drumbeat and, gathering together ministers, Brahmans and nobles, he spoke about the qualities of King Virtue as if to set him above the full moon in the sky. In public, once more he begged the king's pardon and gave him back his kingdom.

'Henceforth my task is the suppression of any rebels who may rise against you; rule your kingdom, with me as your appointed ally.'

This he proclaimed and, passing a sentence on his false adviser, he returned to his own kingdom together with his army. King Virtue, moreover, adorned in splendour under a white parasol, and sitting upon a golden throne with its legs like those of a gazelle, thought over what he had achieved.

'This very success, and the lives of my thousand ministers besides, would not have been secured had I not been steadfast. It was by fortitude alone that I regained the splendour which I lost, and bestowed the gift of life upon my thousand ministers; truly, without ever losing hope, one should persevere, for the rewards of noble deeds bring true prosperity.'

## Prince Wicked

ONCE there was a prince called Wicked who was the son of the reigning king of Benares. The prince was cruel and pitiless, like a trapped snake, and spoke to people only in order to abuse them or strike them. To all, both within and without the palace, he was like grit in their eyes, or like some monster come to eat them.

One day, he went down to the river bank to bathe, together with all his companions. While they were there, a great storm arose and the sky was covered in darkness.

'Take me out to the middle of the stream,' cried he to the servants, 'there bathe me and bring me back again.'

The servants were much puzzled at first what to do, but they took the wicked prince into mid-stream and decided amongst themselves to make an end of him. So, flinging him away in the middle of the current, they let him drift where the waters would take him.

When they returned to the king and were questioned by him, they told him of the storm and of how, in the great darkness that set in, they had lost sight of the prince who must have gone on ahead. Immediately the king ordered a search for the prince, and the four gates of the city were thrown open and men were sent forth, but none of them could find him. He, while the darkness settled around him, had found a tree-trunk in the waters and, as he clung to it in his great fear of drowning, he was swept down the stream.

Now in the banks of that river at a certain spot there dwelt a snake and a water-rat. In their former births these two had been rich merchants living at Benares; he who had become the snake left forty crores of riches buried in the river bank, while he who had become the water-rat had left thirty crores of riches buried at the selfsame place. As the waters rose, they flowed into the dwellings of the snake and the water-rat, and so the two swam furiously out of their holes, across the current, and came upon the very tree-trunk to which the prince was clinging. There they were joined by a parrot who lived on the river bank in a silk-cotton tree now uprooted by the waters, and who, in trying to fly to safety, had been beaten down by the rains. So the four castaways were carried down the river.

Soon the tree-trunk floated to a bend in the river where a hermitage lay, and the hermit heard the cries of the prince in his distress. He determined straightway that his fellow-creature should not perish, so, plunging into the stream, he caught hold of the tree and, with all his mighty strength, pulled it to the bank and set the prince on the shore. He then took the snake and the rat and the parrot and brought them, together with the prince, to the fire that burned in the hearth of his hermitage. First he warmed the animals, then the prince, and then he brought the animals food and after he had done this, he set food before the prince. The prince was annoyed, however, that the animals should be warmed and fed before he himself was looked after, and anger arose in him against the hermit.

The time came, a few days later, when all four had recovered their strength and were ready to depart. First the snake slithered up to the hermit and said:

'Father, you helped me in time of need. If ever you yourself should need help, just come to the river bank at a certain spot and call "Snake". Forty crores of gold which lie buried there shall be yours.'

Next the rat trotted up to the hermit and said:

'Father, you helped me too in time of need. I am not poor either, and all you have to do whenever you need money is to come to the same spot and call "Rat". I shall then give you my thirty crores of gold.'

Then the parrot made his farewell:

'Father, I have no gold or silver to offer, but if ever you are in need of food, come to the land where I dwell and call "Parrot". I and my people will give you many, many fields of rice.'

Lastly the prince came to say good-bye; but he was determined to put the good hermit to death if he should come to visit him, so he said:

'Come to me when I am made king, and I will give you the four things a hermit requires, that is, clothing, food, a dwelling-place and healing herbs.' With these words the prince returned to his country and soon ascended the throne.

As time went on, the hermit decided to test the honesty of his friends, so he went to the spot on the river bank where the snake dwelt and called out 'Snake'. Straightway the snake slipped out of his hole and, bowing deeply to the hermit, said:

'Beneath the ground here are the forty crores of gold. Dig them up and take them away.'

'That is good,' said the hermit. 'I shall not forget in time of need.' Next he called out 'Rat'. Immediately upon his call the rat trotted out of its hole and, resting its whiskers on the ground in deep obeisance, said:

'Down by my home here are thirty crores of gold. Dig them up and take them away.'

The hermit thanked him too and then went to test the parrot. As soon as the parrot heard the hermit calling him, he flew down from his tree-top and asked if he and his parrot friends should gather rice for him from the slopes of the Himalayas.

The parrot also received the hermit's thanks as well as the promise that he would not forget the offer.

Finally the hermit decided to test the prince who had become king and who at that moment was making a tour of his royal city, seated upon a magnificent elephant and followed by the countless members of his court. As soon as the king caught sight of the hermit, he said to himself:

'Here is this fellow coming to ask me for his food and lodging. I must execute him quickly before he can tell everyone how he saved my life.'

He therefore called his attendants, pointed the hermit out to them and, telling them that he thought the man wanted to annoy him, gave them the following orders:

'Seize him and bind him; flog him at every street corner, march him out of the city, chop off his head at the place of execution and impale his body on a stake.'

The servants did as they were told and bound the innocent hermit. They flogged him also as they took him to the place of execution, but all that he could be heard to say was:

'Better to salvage a log than some men.'

Finally some wise men among the bystanders made the king's servants stop for a moment and asked the hermit what he meant by this saying. He told them forthwith the whole story of how he had rescued the prince, as he then was, from the flood. Immediately all the wise men and nobles exclaimed:

'How then shall we be ruled justly by this tyrant if this is how he repays the supreme act of this hermit who has saved his life?'

Without any delay, they rushed on the wicked king with one accord, struck him with swords and threw him by the heels, dead, into the ditch. Then they anointed the hermit to rule over them.

As he ruled over the people he thought he would like once again to try the snake and the rat and the parrot. So one day, followed by a large retinue, he went down to where the snake dwelt and once again he called 'Snake'. Straightway the snake

slipped out of his hole and, bowing deeply to the hermit, said:
'Here is my treasure, O king, please take it.'

The king then told his attendants to take the forty crores of
gold and he proceeded to where the rat dwelt, calling 'Rat'.
The rat came out in the same way and yielded his thirty crores.
When the attendants had taken this treasure as well, the king
went to the tree where the parrot dwelt and called 'Parrot'.
The bird swooped down to the king's feet and asked the king
whether it should bring rice for him.

'The time has not yet come when we need rice,' said the
king. 'Let us all return now.'

So they all returned to the city, the king, the snake, the rat,
the parrot and the seventy crores of gold carried by the attendants. The king had the gold put in a carefully guarded treasure-
house, while for the snake he made a golden tube as a home.
The rat he housed in a crystal casket and the parrot in a cage of
gold. He also gave orders for the three to be fed every day in
vessels of gold, sweet parched-corn for the parrot and the snake,
and scented rice for the rat. In this way, with the king performing all kinds of charities, the four lived their lives in peace and
friendship.

## Suppāraka the Mariner

LONG ago in the seaport of Bharukaccha, in the kingdom of Bharu, a handsome and golden-coloured son was born to the senior navigator. They gave him the name of Suppāraka, and he grew up with great distinction. As soon as he was sixteen years of age he acquired complete mastery in navigation. Subsequently, on his father's death, he became chief mariner and carried on the profession of navigator, being so knowledgeable and full of wisdom that, with him on board, no ship ever came to grief. In the course of time the constant lash of salt water destroyed the sight of both his eyes. From that time on, although he had been chief mariner, he gave up the profession of navigating and approached the king with the idea of gaining a living by serving him. So the king put him to work as an assessor. From then on he assayed the finest elephants and horses, pearls and jewels.

Then one day they brought along to the king an elephant the colour of a black rock, intending that it should serve as an elephant of state for the king. On seeing it the king told them to show it to the wise assessor, and so they brought it along to him. He ran his hand over the elephant's body and pronounced as follows:

'This is not suitable to become an elephant of state since its hind quarters are dwarfish. Evidently, when it was born, its mother was not able to take it on her shoulder with the result that it fell to the ground and became somewhat deformed in its hind legs.'

They took the elephant back to where they had acquired it and asked accordingly. They were then told that the wise assessor had divined the matter correctly. When the king came to know of the matter he was pleased and ordered that eight small coins, a trifling sum, be given to him.

Another day they brought him a horse, thinking that it would serve as one of the king's state horses. He groomed it carefully and then said:

'This is not suitable for a king's state horse for its mother died the day it was born and so, as it did not get its mother's milk, it did not grow properly to full size.'

This verdict also was proved to be true. When the king heard of it he was again pleased and ordered that eight small coins should be sent along.

Then on another day when a chariot was brought with the idea that it would make a state chariot, the king once again ordered that it should be sent to his assessor. He passed his hand over it and then told them that it had been made from the wood of a hollow tree and therefore would not serve as a state chariot. Once more his judgement was found to be correct and the king, being pleased, again ordered that he should be given eight small coins.

Then they brought along a magnificent rug. This also the king had sent to him. On running his hand over it the assessor told them that at one place a hole had been eaten in it by a mouse. They examined it and, on finding that it was so, they informed the king. As before, the king was delighted and commanded that a further eight small coins be given.

The man began to reflect then: After seeing wonders of this kind the king commands that I should be rewarded with eight small coins. Such a fee would be given only to a mere barber and indeed this is just a barber's existence. What is the point of my remaining in the service of such a king? I will go back to my own home. And with these thoughts in mind, he returned to Bharukaccha.

While he was living there, some merchants who had built a ship were wondering whom they should appoint as navigator and it occurred to them that no ship with the wise Suppāraka on board had ever come to disaster, for he was skilled and full of resource; indeed, even though blind, there was none superior to the wise Suppāraka. They therefore approached him and said:

'Be our navigator!'

'My good friends,' said he, 'I am blind, so how can I become your navigator?'

'Even though you are blind,' they answered, 'you still are the best amongst us.'

'Very well then,' he said, after being entreated again and again, 'since you are in earnest, I shall become your navigator,' and so he boarded their ship.

They set sail on the high seas and for seven days proceeded in their ship without mishap. Suddenly an untimely wind arose, and the ship, after scudding along for four months over primeval oceans, came to a sea, Khuramāla by name. There were fish in that sea with the bodies of men and sword-like snouts and they kept shooting out of the sea and diving back in again. When the merchants saw them, they asked their great leader what the name of the ocean was:

'Let's ask Suppāraka; these men with sword-like snout,
Which of the seas is this, where they dive in and out?'

The great leader, on being questioned by them, called to mind his great stock of navigator's lore and answered them as follows:

'You sailed from Bharukaccha, merchants seeking gain;
They call it Khuramāla where your ship has lain.'

In that sea, moreover, there were diamonds to be found. The great leader thought to himself that if he told them it was a diamond sea, then, out of greed, they would gather up a mass of diamonds and capsize the ship. So, saying nothing to them about it, he gave orders for the ship to be hove to and, taking a rope, he let a net over the side as if he were about to catch fish. When he drew it in again, full of sparkling diamonds, he threw them into the hold of the ship and threw overboard some other cargo of little value to make way for them. Then, when the ship had passed over that sea, it sailed into the Aggimāla sea. This sea kept sending out a brilliance like a furiously blazing fire or like the sun at noonday. The merchants therefore asked their navigator:

'We ask Suppāraka: this ocean now in sight,
Which of the seas is this, like sun or fiery light?'

The great leader answered them as follows:

'You sailed from Bharukaccha, merchants seeking gain;
They call it Aggimāla where your ship has lain.'

In this sea now there was gold piled up high. So their great leader in the same way again gathered up gold and put it into the ship. Then the ship, once it sailed beyond that ocean, reached a sea called Dadhimāla which gleamed white, like milk or curds. The merchants therefore asked him:

'Let's ask Suppāraka: this ocean with its gleam:
Which of the seas is this, like milk or curdled cream?'

The great leader informed them again:

'You sailed from Bharukaccha, merchants seeking gain;
They call it Dadhimāla where your ship has lain.'

Silver it was which had collected in this sea. So once again he gathered some up as before and stored it in the ship. When the ship sailed further it then came into a sea called Nīlavannakusamāla which looked like a field of dark *kusha* grass or ripe corn. The merchants asked their leader:

'Let's ask Suppāraka: this ocean we now pass,
Which of the seas is this, like corn or *kusha* grass?'

The great leader told them:

'You sailed from Bharukaccha, merchants seeking gain;
They call it Kusamāla where your ship has lain.'

This was a sea in which hoards of emeralds were to be found. Again in the same way he drew some in and cast them into the ship. Once more the ship sailed on and arrived in an ocean called Nalamāla which looked like a forest of reeds or bamboos. The merchants asked the great leader as before:

'Let's ask Suppāraka: this ocean now in view:
Which of the seas is this, like reed or tall bamboo?'

The great leader informed them:

'You sailed from Bharukaccha, merchants seeking gain;
They call it Nalamāla where your ship has lain.'

In this sea one could find coral the colour of bamboo, so this too he hauled in and put some down in the ship.

When the merchants had passed beyond the Nalamāla, they came to a sea called Valabhāmukha. Here the water is constantly sucked away and rises up again on all sides, and the water which climbs sheer on all sides looks like the cleft walls all round a mighty pit, so that there appears to be a lofty precipice on account of the wave rising up. As well, a terrifying sound breaks out, loud enough to bruise the ears or make the heart faint. When the merchants saw all this, they asked the great leader:

'A dreadful fearful sea, full of inhuman sound;
Let's ask Suppāraka: this ocean we have found,
Which of the seas is this, with pits and cliffs around?'

The great leader, for the last time, told them:

'You have sailed from Bharukaccha, merchants seeking
gain;
Valabhāmukha the sea, where your ship has lain.'

He then said to them, 'My friends, now that the ship has
reached the sea called Valabhāmukha, there is no way for it to
turn back; once this sea gets hold of a ship, it draws it to total
destruction.'

Now seven hundred people had come on board originally
and they were all in great fear of death, so with one accord
they all uttered a pitiful cry like that of souls burning in the
lowest hell. The great leader then thought to himself that,
except for him, nobody else would be able to ensure the safety
of these souls. He therefore determined to save them through
an act of faith, and accordingly gave them the following order:

'My friends, bathe me quickly in scented water and, when
you have clothed me in new clothes, prepare a full bowl of food
and place me in the prow of the ship.'

They rapidly did as they were commanded. The great leader
then took the bowl in both hands, stood in the prow of the ship
and, performing his act of faith, spoke this final verse:

'Since when I can recall, since wisdom came to me,
I know not any being I harmed deliberately;
By this true word, safe may the ship return from sea.'

The ship, which had by then been sailing for four months in
distant seas, returned as if by some miraculous power of truth
in one single day to the seaport of Bharukaccha. Not only did it
arrive at the port, but, soaring over dry land, it reached the
very door of the navigator's home.

The great leader then distributed the gold, silver, emeralds,
coral and diamonds and said, 'My friends, with these you have
enough treasure; travel no more over the seas.' After talking to
them he continued to do worthy deeds his life long, and finally
reached heaven.

## The Prince's Revenge

In the palace of the king of Kosala there was always food
available for five hundred monks, but because there was
never any feeling of comradeship there, the monks pre-
ferred to take their food to the houses of some more friendly
persons and eat with them. A present being brought one day
to the king, he decided that it should be given to the monks
and so he had it sent to the dining-hall. When he was told,
however, that there were no monks in the dining-hall, he
asked:

'Where have they gone?'

'They are sitting down to eat in the houses of their friends,'
was the reply.

So when he had finished his breakfast, he went to the Master
who presided over the monks and asked him:

'What is the best kind of food?'

'The best, great king, is that of friendship; even sour rice-
gruel becomes sweet when offered with the hand of friendship.'

'Then with whom do the monks find friendship?' the king
went on.

'Either with their families or with the families of the Sakyas,
O great king,' the Master replied.

This set the king thinking and he said to himself, 'If I were to marry a daughter of one of the Sakyas and make her my chief queen, then the monks would be friendly with me on account of this relationship.' Whereupon he rose from his chair, returned to his palace and forthwith sent a messenger to Kapilavatthu with this message:

'May you give me a daughter in marriage, for I wish to become related to your family.' When the Sakyas heard the words of the messenger they assembled together and argued amongst themselves as follows:

'We dwell in a territory under the sway of the king of Kosala; if we do not give him one of our daughters there may be some reprisal, whereas if we do give a daughter we shall then be breaking the traditions of our clan, which forbid us to marry with anyone from another clan; so what are we to do?'

Then one called Mahānāma spoke:

'Do not be despondent, for I have a daughter Vāsabhakhattiyā by name whose mother was the slave woman Nāgamundā. The girl is now sixteen years old, extremely beautiful and pleasant natured, as well as being of noble birth on her father's side. Let us send her to him, saying to him, "This is a girl of warrior race."'

The Sakyas approved of this as a good plan and had the messengers sent for.

'We agree,' the Sakyas told them. 'We shall give you a girl whom you may take with you straightway.'

The messengers consulted amongst themselves and concluded:

'The Sakyas are a very proud people on matters of birth; supposing they were to give us a girl whom they claimed to be one of their own, when in fact she might prove to be nothing of the sort. Let us therefore take only a girl whom we see actually eating in company with them.' For they knew that the Sakyas would eat only in the company of someone who was by birth of the same caste as themselves. With this in mind they answered:

'We will certainly take the girl, but we will take with us only a girl whom we have seen eating in your company.'

The Sakyas gave them a place to lodge in and then wondered amongst themselves what they should do.

Mahānāma said: 'Do not be despondent, for I shall find a way out of the difficulty. Dress Vāsabhakhattiyā in her best clothes and bring her along when I am having my meal. Then, as soon as I have taken the first mouthful, you must bring a letter and say to me: "Sir, King Such-and-such sends you a letter; please hear what the message says."'

They agreed to what he had to say and, as he was taking his meal, they dressed the girl in the finest clothes.

Mahānāma then said: 'Bring my daughter along and let her take food with me.'

They told him that she was just being adorned in her finery and, after a moment's delay, they brought her in. The girl, imagining that she was about to eat with her father, put her hand into the same dish. Mahānāma took a piece of food at the same time and put it into his mouth, but barely had he reached out his hand for a second mouthful than the letter was brought in to him and he was told: 'Sir, a certain king has sent you a letter; please hear what the message says.'

Mahānāma stayed his right hand in the dish, holding some food, while with his left he took the letter and studied it. While he went on studying the letter, the girl continued with her meal, and, as she finished, he wiped his hand and his mouth clean. The messengers all came definitely to the conclusion that she was his daughter; not for a moment did they suspect anything further.

Mahānāma therefore sent his daughter off with a great retinue, and when the messengers brought her to Sāvatthi they announced:

'This girl is Mahānāma's daughter by birth.' The king was greatly pleased and had all the city gaily bedecked. He placed her on a heap of jewels and had her consecrated as his chief queen, and she was dear to him and beloved by him. Soon a son was born to them which was looked after in royal fashion. The colour of the child was a golden brown and when the day came for him to be named, the king sent a message to his grandmother, saying:

'Vāsabhakhattiyā, the daughter of the Sakya king, has given me a son. What name shall he be given?'

The minister who took the message was slightly deaf but he went along to the grandmother and informed her. When she had heard him she said:

'Even without a son Vāsabhakhattiyā was more important than anyone else, but now she will be very dear to the king indeed.'

The deaf minister heard the words 'very dear' incorrectly and, thinking that it was 'Vidūdabha', he went to the king and informed him:

'Sir, you are to name the child "Vidūdabha".'

The king thought that this must be some old family name and accordingly called him Vidūdabha.

The young prince grew up, being treated with every royal care. When he reached the age of seven years he noticed how the other young princes received presents from their mothers' fathers of toy elephants, horses and the like, and so he asked his mother:

'Mother, presents come to the others from their mothers' parents, so why do mine never send me anything? Have you no parents?'

She answered him, without telling him the whole truth, by saying:

'My dear, your grandparents are the Sakya kings and they live a long way off; that is why they send you nothing!'

Again when he was sixteen years old he said to her:

'Mother, I would like to see my grandparents.'

She put him off by saying:

'Enough, my child, what would you do there anyway?' but he kept asking her again and again.

At last his mother said, 'Very well, go then.' He asked his

father's permission as well and then set out with a mighty retinue. Vāsabhakhattiyā, however, had a letter sent on before him in which she said that she was living happily where she was and that they were not to tell the boy anything further. When they heard about the arrival of Vidūdabha the Sakyas decided that they could not receive him properly and accordingly sent all their young children off to the country. When the young prince arrived at Kapilavatthu, the Sakyas all assembled in the council hall. The prince went up to the council hall and stopped there. Then they said to him in order to introduce him:

'This is your mother's father, this is her brother.'

He went round them all paying his respects to each one. Yet even though he bowed to them all until his back was nearly broken, he saw that not one of them was greeting him in return:

'Why is it that not one of you greets me?' he asked.

The Sakyas answered him, 'Dear fellow, all our young princes have gone to the country,' and with that they proceeded to show him great hospitality.

After he had stayed there for a few days he set forth again with his vast retinue. At that very moment a female slave was cleaning down with milk and water the bench in the council hall upon which the prince had sat, and as she did so, she complained loudly: 'This is the bench where the son of the slave Vāsabhakhattiyā sat.'

One man from the prince's retinue who had forgotten his spear went back to get it and heard the insults about Prince Vidūdabha. He inquired further from the woman and learned that the prince's grandparents were the Sakya Mahānāma and the slave Nāgamundā. On his return he told his fellow-soldiers, whereupon a great clamour arose:

'Vāsabhakhattiyā is the daughter of a slave woman.'

The prince heard it and a bitter thought arose in his mind: Let them wash the bench now where I sat with milk and water; when I have come to the throne I will take the blood from their throats and wash the place where I sit with that.

When he returned to Sāvatthi, the ministers told the whole story to the king. The king, in his fury, said:

'So, they have given me the daughter of a slave woman to wife.' And straightway he cut down the daily allowance given to Vāsabhakhattiyā and her son, leaving them only what slave men and women were allowed to have.

Then, a few days later, the Master came to the king's palace and sat himself down. The king came in and greeted him, saying:

'Sir, your relations have given me the daughter of a slave woman to wife, so I have cut down the daily allowance given to her and her son, leaving them only what slave men and women are allowed to have.'

'The Sakyas did wrong indeed, great king,' the Master replied, 'for they should have given a girl of equal rank, but this much I say to you, great king: Vāsabhakhattiyā, as the daughter of a king, was consecrated in the house of a noble king, while Vidūdabha was born the offspring of a noble king. The wise men of old knew this when they said that the birth of the mother was of no account but that the birth of the father it was that mattered, for they gave the position of chief queen to a poor woman, a gatherer of wood. Her son became king and acquired the sovereignty over Benares, twelve leagues in extent.'

The king, when he heard this just story, said:

'Indeed it is the birth of the father which matters,' and, being pleased, once again gave the mother and her son their proper allowance.

In due course the king died and Vidūdabha came to the throne. Remembering his old enmity he set out with a mighty army fully determined to kill the Sakyas. It so happened at dawn that day that the Master was looking out over the world and he foresaw the annihilation of the Sakyas. 'I must do something to help my people,' he thought, and so in the morning he went seeking alms, and after he had returned from his meal, he lay down, lion-like, in his room which was perfumed with incense. In the afternoon he went outside and sat down at the foot of a tree which gave but scanty shade, on the outskirts of Kapilavatthu. There stood also, not very far from there, a vast *banyan* tree which gave thick shade, by the boundary of Vidūdabha's realms. When Vidūdabha saw the Master, he went up to him and greeted him:

'Sir, why do you sit under this poor tree in the heat of the day? Pray sit at the foot of this shady *banyan* tree.'

'Let it be so, great king, for it is the shade of my people which keeps me cool,' the Master replied. Vidūdabha realized that the Master must have come to protect his people; he therefore bid him farewell and returned to Sāvatthi. The Master then got up and went to Jetavana monastery.

A second time the king remembered his grudge against the Sakyas and again set forth. Again he saw the Master and again he returned. A third and yet a fourth time he set out, and the Master, realizing that nothing would be done to avoid the evil deeds of the Sakyas in casting poison into the river, did not appear the fourth time. Vidūdabha then slew all the Sakyas, even to the babes in arms, and once he had washed the bench where he had sat with the blood from their throats, he returned to his kingdom.

# The King's Remorse

ONCE there was a king of Kosala who had a commander in chief named Bandhula. Bandhula and his wife Mallikā had no children, so he sent her back to her home town, saying to her, 'Return to your own family.' She answered him that she would do this but that first she must see the Master of the monastery. Accordingly she entered Jetavana monastery and greeted the Master. As she stood on one side he asked her:

'Why have you come?' and she said to him, 'My husband has sent me back to my people's house.'

'Why has he done this?' the Master asked.

'Because we have no children,' she replied.

'If that is all that is wrong,' the Master went on, 'there is no need for you to go; return to your own home.'

Mallikā was delighted and, after bidding the Master farewell, she returned to her husband's roof. When Bandhula asked her why she had returned she told him that the Master had sent her back.

'Well,' said he, 'he alone must have seen the reason for this.'

Not long afterwards she knew that they would soon have a child and she begged her husband to take her to the tank in the city of Vesāli where the royal families of Licchavis drew water for the anointing of kings, for there, said she, she wanted to bathe and drink. Her husband, the commander in chief, promised that he would do this and, taking his bow which was as strong as a thousand bows, he placed her in his chariot. He set forth from Sāvatthī and driving onwards in his chariot he entered Vesāli.

At that time there happened to be a blind Licchavi, Mahāli by name, who lived near the gate of the town. He had been educated in the same school as the king of Kosala's commander in chief, Bandhula, and he was occupied with instructing the Licchavis in matters of policy and justice. When he heard the thundering of the chariot past the city limits, he said:

'There is the noise of Bandhula's chariot; today a danger arises for the Licchavis.'

A strong guard had been set both outside and inside the precincts of the tank, while above it a wire net had been stretched so that there was not space for even birds to fly through. The commander in chief, however, got down from his chariot and laid about him with his sword amongst the guards, putting them to flight. He cut through the iron net and let his wife bathe and drink in the tank. Then he himself bathed and when he had set Mallikā once more in the chariot, he set off home the way he had come. The kings of the Licchavis were greatly angered and five hundred of them mounted five hundred chariots and set off in pursuit, determined to capture Bandhula. They told Mahāli what had happened, and he warned them not to go, for, said he: 'Bandhula will kill you all.'

They told him that they would go nonetheless.

'In that case,' he replied, 'if you see a place where a wheel has sunk down to the axle, you should return; if you return not then and you hear a sound in front of you like that of a thunderbolt, return from that place; if even then you do not return and you see a hole in the front of your chariots, then return, nor go farther than that place.'

They did not turn back, however, at his word but kept on in their pursuit.

Mallikā saw them and said: 'There are chariots in sight, my lord.'

'Then tell me when they all look as if they were one chariot,' was her husband's answer.

As soon as they all appeared in a line as one chariot she spoke again: 'Now, my lord, the front of only one chariot can be seen.' Bandhula then said to her: 'Gather up the reins,' and, giving her the reins, he stood up in the chariot and made ready his bow.

The wheel of the chariot sank down to the axle, but when the Licchavis came and saw that place they did not turn back. When Bandhula had gone a little further, he slapped the bowstring and a sound as of a thunderbolt was heard, but even then the Licchavis did not turn back, keeping on all the time with their pursuit. Then Bandhula, standing up in his chariot, let fly an arrow, and the arrow made a hole in the foremost of the five hundred chariots and after going on to pierce through the five hundred kings where their girdles were fastened, it entered into the earth.

They, however, did not realize that they had been pierced and they kept on pursuing, shouting: 'Stop, stop there!'

Bandhula stopped his chariot and said to them: 'You are dead men and I cannot fight with dead men.'

'Are then dead men supposed to be like us?' they asked.

'In that case, just undo the girdle of the first man,' he replied.

They undid it, and barely was his girdle loosened than he fell down dead.

Bandhula went on to say to them: 'You are all in the same condition, so go back to your own homes and set in order whatever must be set in order; make your will known to your sons and wives and then take off your armour.'

They did so accordingly and thereupon they all died.

Bandhula then brought Mallikā back to Sāvatthi and in the course of time they had sixteen pairs of twins. All of them grew up to be heroes endowed with strength and they all became skilled in every art. Each one had a retinue of a thousand men and when, together with their father, they went to the palace of the king, they filled the courtyard of the king to overflowing.

Then one day some men who had been defeated on a false charge in court saw Bandhula coming and raised a great clamour. They told him that the judges of the court were the cause of the charge being falsely put. Bandhula therefore went into the court and, judging the case himself, made a fitting award to each man. The vast crowd of people sent up a great shout of applause. The king asked what had happened and, on hearing of the matter, was greatly pleased. He then deposed the judges and assigned the judgement of cases to Bandhula, who from then onwards administered justice.

The former judges then, no longer receiving their bribes, became poor and they began to plot against Bandhula in the king's palace, saying that he was seeking the kingdom for himself. When the king heard their story he was not able to put the thought out of his head.

If he is slain on the spot here, the blame will fall on me, he thought, so, after pondering over the matter, he hired some soldiers and commanded them to attack the frontier. Then he sent for Bandhula and said to him:

'The frontier is in a turmoil; go together with your sons and capture the villains.'

After he had given him these instructions, he sent with him some further mighty warriors, commanding them to cut off the heads of Bandhula and his thirty-two sons and to bring them back. When Bandhula reached the frontier, the mercenaries all fled as soon as they knew that the commander in chief was coming. He settled that province duly and, once he had quietened the district, he returned home. When he was only a short way from the city, the soldiers cut off his head and those of his sons.

It happened on that day that Mallikā had invited two chief disciples together with five hundred monks. In the morning a letter was brought and given to her which said:

'The heads of your husband and your sons have been cut off.'

When she heard the news she did not say anything to anyone but put the letter in her dress and went on to entertain the company of brethren. Her attendants were giving the monks food and as they brought in a bowl of *ghee*, a kind of clarified butter, they broke the bowl in front of the elders. The chief amongst them said:

'What is liable to be broken has been broken; do not give it a thought.'

She then drew the letter from her dress and said to them:

'They have brought me this letter saying that the heads of my husband and of my thirty-two sons have been cut off, so that if I do not worry at having heard that news, am I likely to worry about a pot being broken?'

The leader of the elders rose from his seat and went home. She then called together her thirty-two daughters-in-law and spoke to them as follows:

'Your husbands, though they have not sinned, have reaped the fruit of deeds done in a former birth, but you are not to grieve nor bear any anger in your hearts against the king.'

When the king's spies heard this they went and told the king how the women bore no grudge. The king was greatly disturbed and, going to her house, he asked forgiveness of Mallikā and her daughters-in-law and granted them a wish. She told him that she would accept and when he had gone she made an offering of food to her ancestral gods, and bathed. She then approached the king and said to him:

'Your majesty has granted me a wish, but I ask nothing more than this: please allow me and my thirty-two daughters-in-law to return to the homes of our own families.'

The king forthwith granted his permission, so she then sent each one of her daughters-in-law to her own home and she herself returned to her family home. The king, moreover, gave the post of commander in chief to one Dīghakārāyana, the son of Bandhula's sister, but Dīghakārāyana went around constantly thinking about how the king had murdered his uncle, and so he bided his time.

From the time of the murder of the innocent Bandhula, the king was full of remorse nor did he gain any peace of mind nor find gladness in being king. At that time the Master was living near a country town called Ulumpa, belonging to the Sakyas. The king went there, taking with him his son, the young prince, and pitched camp not far from the park. Then, after giving to Dīghakārāyana the five symbols of royalty, the yak-tail fan, the turban, the sword, the umbrella and the sandals, he went with a small following to the monastery, in order to salute the Master, and alone he entered the room perfumed with incense.

After he had entered the room fragrant with incense, Dīghakārāyana took the symbols of royalty and, making the young prince king in his stead, he then returned to Sāvatthi, leaving for the king just one horse and one serving woman.

When the king had talked for a while in a friendly way with the Master he came out again but saw no army. He asked the serving woman what had happened and on hearing the story he set off in haste, solely bent on capturing Dīghakārāyana for taking the young prince with him. When he reached the town it was late and the gates were shut, so he lay down under a shelter, wearied by the wind and the sun, and there during the night he died. As morning came, people heard the sound of the woman as she wailed, 'Sir, the king of Kosala has lost his life,' and they informed the new king. The latter then performed the funeral rites for his father with all due worship.

# King Shibi

THERE was formerly a king named Shibi who was self-denying, highly compassionate, generous and resolute and who granted safety to all creatures. In an attempt to confound him the great god Indra once changed himself into the form of a hawk and swiftly pursued *Dharma*, the nature of law and righteousness, which by magic had turned into a dove. The dove took flight in fear and sought refuge in Shibi's lap, whereupon the hawk then spoke to that king with human voice:

'O king, this is my food, give the dove to me who am hungry, for otherwise I shall die and then where will be your righteousness?'

Then Shibi spoke to him: 'This creature has come to me for shelter and will not be forsaken by me, but I will give you other flesh which is the same as this.'

The hawk said, 'If that is so, then give me your own flesh.'

That king in his delight agreed, saying, 'Let it be so.'

But as the king cut off pieces of his own flesh, placing them on the scales, so the dove became the heavier in the balance. Then the king placed his whole body in the scales, and a voice came from heaven, saying: 'Well done, well done, that is indeed equal.'

Indra and *Dharma* then forsook their forms as hawk and dove and in their delight made King Shibi's body whole again and when they had granted him many other blessings, they vanished away.

## A Nasty Rumour

THERE is, beside the Ganges, a city named Kusumapura, and in that city there lived an ascetic named Harasvāmin who was fond of holy places. Being an ascetic, he lived a religious life, without possessions and devoting himself to meditation. He had been born a Brahman and so was of the highest caste; and as a Brahman he gained his living by begging, becoming the object of reverence among the people on account of the excellence of his penance. Then one time as he went out begging, a wretch who begrudged his good qualities saw him in the distance and called out from the middle of the crowd:

'Do you not know what kind of deceitful ascetic we have here? He it is who has eaten up all the children in the town.'

A second fellow of the same ilk, when he heard this, followed up by saying:

'True enough, I have heard people saying just the same thing.'

'It is a fact,' said a third also, backing him up, for the chain of evil gossip binds blame upon the good.

In this way the rumour went round gradually from ear to ear and was believed by everyone throughout the city. And all the citizens forcibly restrained their children within their houses, thinking that Harasvāmin would take the young ones away and eat them all. Then the Brahmans there assembled together, fearful lest their children should be destroyed, and discussed banishing him from the town. As they did not dare speak openly about it for fear that he might get angry and gobble them up, they dispatched messengers instead. The messengers went off then and addressed him from a distance:

'The Brahmans say that you must leave this city.'

At these words the astonished man asked: 'For what reason?' and they answered him:

'You are eating the children here on sight.'

When Harasvāmin heard this he went to visit those Brahmans, wanting to reassure them and wishing to allay their fear. And when the Brahmans saw him coming they climbed in their terror high into the temple, for people who are befuddled by rumour are not generally capable of discriminating. Then Harasvāmin, standing below those Brahmans, called each one of them by name and spoke to them as they stood above him:

'What now is this fanciful idea of yours, O Brahmans; why do you not consult among yourselves as to how many children I have eaten, and whose children and how many of each man's?'

As soon as they heard this, those Brahmans began to inquire one of another, and it transpired that every single child of every one of them was still alive. In course of time other citizens who were appointed to investigate confirmed that the matter was correct, whereupon all the people, Brahmans and merchants, declared:

'Truly this worthy man has been falsely accused by us fools; everybody's children are alive, so whose has he eaten?'

When all of them had spoken to this effect Harasvāmin started out then to leave the city, his innocence proved; for what pleasure does a wise man, whose heart has been disgusted at the blame spread by wicked people, find in an evil place where he is falsely judged? But then, when he was entreated by the Brahmans and merchants bowing at his feet, Harasvāmin somewhat reluctantly agreed to remain there.

## An Indian Love Story

THERE dwelt once in the town of Shrāvastī a man named Shūrasena who, being a connexion of the ruling family, was a Rajput, and his estate consisted of a village. He was in the service of the king and he had a wife suitable to his station from the region of Mālava. Her name was Sushenā and she was dearer to him than life. One day Shūrasena was summoned by the king and was preparing to go to the royal camp when his fond wife said to him:

'My husband, you should not go leaving me all alone here, for I shall not be able to stay here even a short while without you.'

As his beloved spoke to him thus, Shūrasena answered her, 'How can I refuse to go when summoned by the king? Do you not understand this, my dear? As a Rajput I am a servant and dependent for my livelihood upon another.'

When she heard this his wife said to him with tears in her eyes:

'If you definitely must go, then I will bear it somehow or other, as long as you come back not a day beyond the beginning of spring.'

When he heard this he said to her finally:

'Very well, my dear, even if I abandon my post, I shall come back on the first day in the month of Chaitra.'

With these words Shūrasena managed somehow to take leave of his wife and went to the royal camp to attend upon the king. His wife, for her part, remained there counting the days and waiting for that day at the beginning of the spring which would herald her husband's arrival. Then as the days went by, there came the day of the spring festival and the sound of the cuckoos joyfully announcing their summons to the god of love. One could hear too the hum of the bees drunk with the fragrance of flowers like the sound of the bow being strung by Kāma, god of love.

Then one day the wife of Shūrasena reflected to herself: Now has come the day of the spring festival and assuredly my husband will return today. Then Sushenā bathed herself and paid her devotions to the god of love; adorning herself she kept watching out for her husband to arrive. But as the day went on and he did not come, she began to think to herself, sorrowing and without hope in the night: Now that my loved one has not come back to me, the time for my death has indeed come; where indeed can those who are fully intent on serving a stranger find love for their own family? And as she pondered these thoughts with her heart fixed on her husband, her life left her, as if burnt up in the wild fire of the god of love.

At the same time Shūrasena managed somehow to obtain leave from the king just that very day, and, eager to see his loved one, he mounted a magnificent camel and, passing over the long distance, he reached his own home in the last watch of the night. There he saw his dear one, dead, arrayed with her ornaments, like a flower with its blossoms full-blown, uprooted by the wind. As he looked at her he was distressed beyond measure and as he took her in his arms with lamentations, his life too left him on that instant.

When Chandī, the granter of blessings, saw the fate that the pair had suffered, she as the goddess of families, in her compassion brought them to life again. Then the pair, on receiving their lives again, each having seen the other's affection, lived together inseparable from that time on.

## Mousey the Merchant

I N a certain city one day in the midst of some merchants who
were discussing amongst themselves their skill in their respective trades, one merchant spoke as follows:

'It is no great wonder that a man who is economical with
his riches should acquire riches, for once, though I was without
money, I gained a fortune. My father died before I was born,
and wicked relatives then took everything from my mother so
that, out of fear of them and for the safety of her unborn child,
she went away and stayed in the house of Kumāradatta, a
friend of my father. There I was born to that worthy woman to
be the sole means of her support and she, by performing menial
tasks, brought me up there. Then she, though being so poor,
persuaded a teacher to take me on so that gradually I was
taught writing and arithmetic. Then my mother said to me:

' "You are the son of a merchant, so now, my son, you must
engage in trade; there is a very rich merchant in this district
called Vishākhila. He gives stock on credit to those of good
families who are distressed, so go and ask him for credit."

'Thereupon I went to see him, but at that very moment
Vishākhila was talking angrily to some merchant's son:

' "Even with this mouse which you see lying dead here on

the floor as stock in trade, a clever fellow would be able to make money; I gave you many coins, yet, so far from gaining interest, you have not even kept the capital intact."

'When I heard this, I quickly said to Vishākhila:

'"I am taking this mouse from you as stock on credit," and with these words I picked the mouse up and entering it in his ledger I went off, while, for his part, the merchant burst out laughing. Then, in exchange for two handfuls of chickpeas, I gave the mouse to a certain merchant for his cat. I then ground those peas and, taking a jar of water, I went and stood in a shady spot outside the town by the cross-roads. There I offered the cool water and the peas with all civility to a band of wood-cutters who were returning home exhausted. Out of gratitude each of those woodcutters gave me two pieces of wood, where-upon I brought that wood to the market and sold it. Then with the little bit of money I bought more peas and on the next day in the same way I received wood from the woodcutters. By doing this every day and thus gradually acquiring capital, I bought the entire stock of wood for three days from the wood-cutters.

'By chance then there arose a shortage of wood due to excessive rains and so I sold all that wood for many hundreds of coins. In this way I set up shop with the money and by carrying on trade I have gradually become very rich through my own skill. I then made a mouse out of gold and sent it to that Vishākhila, whereupon he gave me his daughter in marriage. This is the reason why people call me "Mousey", for I have gained prosperity though I was once without any money.'

When they heard this all the other merchants were amazed; for how would the mind fail to be surprised at a picture painted in thin air?

## Riches or Happiness?

THERE was formerly in this country a city whose king had a young attendant of the Kshatriya, or warrior, caste called Yashovarman. Yet, although the king was generous he never gave him anything whatsoever. Whenever the king was asked by him in his distress for anything, he would invariably point out the sun and say:

'I would like to give you something, but this worshipful god does not allow me to give you anything; so what am I to do? Tell me!'

So the other remained there in misery while he waited for his opportunity, and then there came the moment for an eclipse of the sun. At that moment Yashovarman went, always in his role as attendant, and informed the king who was busy giving away many great gifts: 'He who does not allow you to give anything to me, that sun, O master, has now been swallowed by his enemy, so just give me something.' When the king heard this he laughed, and, having already given away many presents, he gave him clothes and gold and so forth.

As the riches were gradually consumed, Yashovarman became despondent, for his master gave him nothing further, so, since his wife had died, he went to Durgā, the goddess dwelling in the Vindhya mountains. 'What does it profit me to live with a body as worthless as though it were dead?' he said. 'I shall abandon it in the presence of the goddess or else I will obtain what I desire.' Whereupon he lay down on the strewn *kusha* grass in front of the goddess and with his mind intent on her he performed a great penance by abstaining from all food. Then the goddess showed herself to him in a dream, saying:

'I am pleased with you, my son, and I will give you the blessing of wealth or else that of enjoyment; say which you will have.'

When he heard this, Yashovarman answered the goddess: 'I do not know exactly the difference between these two blessings.'

Then the goddess answered him: 'Go to your own country and visit the two merchants there called Bhogavarman and Arthavarman and inquire into their fortune. Whichever one then will please you better, come here and ask for that.'

After Yashovarman had heard this he awoke the next morning and, having broken his fast, he then went to Kautukapura in his own country.

There he visited first the house of Arthavarman, whose wealth consisted of countless gold, jewels and so forth and had been gained through trading. When he saw his riches he approached him courteously and being welcomed as a guest was invited by him to eat. So then at the side of Arthavarman he enjoyed food that was fitting for a guest, curries of meat and *ghee*. Arthavarman, however, just ate barley meal mixed with a half-measure of *ghee* and a very little quantity of rice and meat curry.

'Merchant, why do you eat so little?' Yashovarman asked the merchant out of curiosity and the latter replied, 'Today out of regard for you I have eaten this small amount of rice with meat curry, also a half-measure of *ghee* and some barley meal. Normally, however, I never eat more than some barley meal with a trace of *ghee;* more than this I cannot digest on account of a weak stomach.'

When Yashovarman heard this he pondered over it and thought but poorly within himself of Arthavarman's barren fortune. Then, when it came on to night-time, the merchant Arthavarman again had milk and rice brought for Yashovarman, but while Yashovarman again ate to his heart's content, Arthavarman took only a small measure of milk. Then the two of them lay down on their beds in the same room and so Yashovarman and Arthavarman gradually fell asleep.

Then in the night Yashovarman saw in a dream some men of terrible appearance with sticks in their hands brazenly entering the room. Angrily they said: 'Wretch! Why did you today eat a tot of *ghee* more than was allowed, as well as meat curry, and why did you drink a measure of milk besides?' Whereupon the men dragged Arthavarman out by the feet and beat him with their cudgels and took out of his stomach the rice, meat, milk and measure of *ghee* which he had consumed over the allowance. When Yashovarman saw this he woke up, looking about him, and there was Arthavarman awake and visited by acute pain. Then, crying out while his stomach was being massaged by his servants, Arthavarman vomited everything which he had eaten in excess. When his pain had at last subsided, Yashovarman thought to himself: 'To the devil with this fortune of riches! What enjoyment of it is there? Anything like this presented as an offering would be the misfortune of destruction.' With thoughts like these in his mind he passed the rest of the night.

On the next morning Yashovarman took his leave of Arthavarman and went to the house of the merchant Bhogavarman. There he greeted him courteously and was invited with all respect by the merchant to eat with him that day. He did not, however, see any accumulation of riches belonging to the merchant, though he saw that his house was beautiful and that he had clothes and ornaments. Then as Yashovarman stood at his side, the merchant Bhogavarman proceeded to carry on his business in his own way. The goods which he had taken in, he quickly gave to another so that without any money of his own

he made money out of the transaction; and then he straightway
sent that money off to his wife by a servant so that she could
gather together all kinds of food and drink. Thereupon a friend
came quickly up to Bhogavarman and said to him: 'Our meal
is ready; get up, come along and let us eat! For the rest of our
friends are there and are waiting to see you.'

But as he spoke this his friend said to him: 'Today I shall not
go, for there is a guest of mine here.'

'Is he not then our friend as well? Get up straightway!' So
Bhogavarman was led obstinately by his friend together with
Yashovarman and they ate the very best of meals. After drinking wine, Bhogavarman came back to his own home in the
evening with Yashovarman and they had a meal of various
kinds of food and drink. When night came he asked his servants:
'Is there wine enough for us now for the rest of the evening?'
When they answered: 'Sir, there is not!' the merchant betook
himself to his bed saying: 'How can we drink water for the rest
of the night?'

Then Yashovarman, sleeping at his side, saw in his sleep a
couple of men entering with others following behind them.
The latter, who were carrying sticks, said to those who had
entered first: 'Why did you not think of wine for Bhogavarman
for the latter part of the evening? What have you been up to,
you wretches?'

With that they angrily belaboured them with blows of their
sticks. Then the men who were being beaten with the sticks
called out: 'Let us be forgiven this one mistake!' Whereupon
the others went out again. When Yashovarman saw this he
woke up and thought: The fortune of enjoyment which
Bhogavarman has, and which falls unexpectedly to one's share,
is commendable indeed, but the fortune of riches belonging to
Arthavarman, even though vast, was bereft of enjoyment.
With these thoughts he passed the rest of the night.

The next morning Yashovarman took his leave of the good
merchant and went once again to the feet of Durgā, the dweller
in the Vindhya mountains. He reminded Durgā of those two

fortunes which she had described to him when he came to her as a penitent on the first occasion, and of those two he chose the fortune of enjoyment. Durgā granted it to him, and then Yashovarman returned to his own home. Thanks to the favour of the goddess, he remained there enjoying happiness on account of his fortune of enjoyment which appeared without even the need for him to give it a thought. In this way a large fortune with no enjoyment, wise men know, is worth far less than a fortune full of enjoyment, however small it be.

## *A Slip of the Tongue*

THERE was a certain poor and stupid Brahman named Harisharman in a certain village and he was in sore straits for want of employment, as well as having very many children as the result of misdeeds in a former birth. As he wandered about with his family getting alms he came in due course to a city and entered into the service of a very rich householder named Sthūladatta. He made his sons cowherds, while his wife performed menial tasks and he himself dwelt near the house carrying out messages.

One day the wedding of Sthūladatta's daughter took place and was attended by numerous relatives and crowded by throngs of processions. On this occasion in the household Harisharman was counting on being able to eat his fill of *ghee* and meat and other things together with his family. Yet while he was hoping for this event, nobody seemed to remember him, whereupon failing to get any food, he spoke to his wife one night:

'Such a lack of importance on my part here is due to my poverty and stupidity; this being the case, I must pretend to be intelligent by means of some trick or other and so make myself worthy of Sthūladatta's respect. When a chance arises, therefore, you must tell him that I am a knowledgeable person.'

This he said to her and then, thinking over the matter in his mind, while everyone was asleep, he took from Sthūladatta's house the horse upon which the son-in-law rode. He then concealed it, tethering it some distance away, and on the next morning although the bridegroom's family searched all over the place they did not find the horse. Then Harisharman's wife approached Sthūladatta who was disturbed at the misfortune and was seeking the horse-thief and said to him:

'My husband is very wise and skilled in sciences such as astronomy; he will get your horse for you; why do you not consult him?'

When Sthūladatta heard this, Harisharman was summoned and said to him:

'Yesterday I was overlooked, but today, now that the horse has been stolen, I am remembered.'

Sthūladatta then placated the Brahman, saying to him:

'Forgive us the omission, but please tell us who has taken our horse away.'

Harisharman then made a pretence of tracing out some lines and said:

'The thieves have tethered the horse on the boundary south from here; it is standing hidden and will not be taken away until the end of the day, so go quickly and bring the steed back.'

When they heard that, several men ran off straightway and finding the horse they brought it back, praising Harisharman's knowledge. After that everyone paid respect to Harisharman as a knowledgeable fellow and he lived comfortably, being honoured by Sthūladatta.

As the days passed by, a considerable amount of treasure consisting of gold and jewels and other things was stolen somehow or other by a thief from the king's palace. When the thief was not discovered, the king, who was well known for his wisdom, had Harisharman speedily summoned. He, when he was brought along, began playing for time and, being frantic as to how he might find out, said: 'I will tell you in the morning.' He was placed in a bedroom by the king and carefully guarded.

Now there was a maidservant in the king's household called Jihvā, which means 'tongue', and it was she, together with her brother, who had taken away the treasure from inside the palace. She was very worried about Harisharman's knowledge and so, in order to find out what she could, she went along in the night to his room and put her ear to the door. At that moment Harisharman, all alone inside, was blaming his own tongue for bragging about his pretended knowledge:

'O tongue, what have you done out of sheer greed for enjoyment? Wicked wretch, you must now endure imprisonment here.'

When the maidservant Jihvā heard this, she thought in terror that she had been found out by the knowledgeable fellow, so by means of a trick she got inside the room to him and falling at his feet she said to that supposedly wise man:

'O Brahman, I am that Jihvā, the Tongue, known by you to have taken the treasure; when I took it, I buried the money in the ground in the garden at the back of the building here underneath a pomegranate tree. Therefore protect me and take this piece of gold which I have in my hand.'

When he heard this, Harisharman haughtily said to her:

'Go, I know all, past and future as well as present; I will not, however, expose you to punishment since you have come to me pitiably seeking refuge, but you must return to me everything that you have kept.'

The maidservant agreed speedily to what had been said to her and swiftly took her leave. Harisharman then thought to himself in astonishment: Fate, being favourable, has accomplished, through the merest whim, a deed which seemed impossible; for here, when disaster was at hand, the affair has been settled unexpectedly for me. The thief Jihvā has fallen down in front of me while I was blaming my own tongue.

Hidden misdeeds come to light, it seems, through guilty feelings. So pondering like this he passed the night very pleased with himself.

The next morning putting on a show of assumed wisdom, he led the king into that garden and got out for him the treasure which had been buried there, but told him that the thief had escaped, taking part of it with him. At this the king was delighted and proceeded to give him many villages.

Then a minister confided in the ear of the king:

'How could there be such wisdom, unattainable by mortals, without learning? Surely therefore this fellow carries on a rogue's existence by being in league with thieves, so let him be tested once more by some trick or other.'

The king therefore casually had a new jar brought which was covered over and which had a frog inside it and he said to Harisharman:

'If you can find out what is inside this jar, I will do you very great honour this day.'

When he heard this he thought his hour of destruction to be at hand, but he remembered his own nickname 'Frog' which his father had playfully given him in his childhood and he began to lament about himself:

'Truly indeed, O Frog, this jar has appeared unexpectedly here to destroy you in your weakness.'

When the people heard this and assumed he meant the object which had been presented to him, they shouted out with cries of 'Lo, the great sage! He has found out the frog!' The king then, being pleased with Harisharman and thinking he had remarkable insight, gave him villages together with gold, wagons and the umbrella as symbol of power, and straightway Harisharman was turned into a regular prince.

## The Mouse and his Friends

THERE was a great *shālmali* tree in a certain spot in a
forest and a crow named Laghupātin lived there having
made his abode in it. One day as he was sitting in his
nest he saw a man who had come to the foot of the tree; he
had a net in his hand and a stick and he was of terrible appearance. No sooner then had the crow noticed him than the
latter stretched his net out on the ground, scattered some grains
of rice there and forthwith concealed himself. At that very
moment the lord of the pigeons, Chitragrīva by name, arrived
there, wandering through the sky surrounded by hundreds of
pigeons. When he saw the strew of rice he fell into the net out
of his desire for food and together with his retinue he was caught
in the meshes of the net. When he saw this, Chitragrīva said to
all his followers:

'Take the net in your beaks and fly quickly up into the air.'

Thereupon all the pigeons said: 'Yes we will!' and taking the
net they flew up swiftly, and in terror began to journey through
the sky. The hunter then arose and came forward, looking up
in distress, whereupon Chitragrīva fearlessly spoke to his followers: 'Let us go quickly to the presence of my friend
Hiranyaka the mouse; he will cut through these bonds and

release us.' After saying this he went together with his followers dragging the net and, approaching the door of the mouse's hole, he descended from the sky.

'Ho there, noble Hiranyaka, come out! It is I, Chitragrīva, who am come.' So the emperor of the pigeons summoned the mouse there. When the mouse heard this and noticed through his doorway that his friend had arrived, he came out from his lair with its hundred openings. When he had come up and inquired about the affair, the friendly mouse eagerly cut through the bonds of the lord of the pigeons and his followers. Chitragrīva then fêted the mouse with affectionate words for having cut the snare and, flying up into the sky, he went off with his followers.

The crow Laghupātin had come along too and had seen all this, so, approaching the door, he addressed the mouse who had gone back into his hole: 'I am the crow Laghupātin; I have seen your tender friendship and I choose you as a friend for being able to be a rescuer from such misfortune!'

When the mouse heard this and saw the crow from within his hole, he said, 'Go away! What friendship could there be between the eater and the eaten?'

'Peace be with you!' the crow replied. 'If I were to eat you I would have but momentary satisfaction, while as a friend your life is constantly protected.' Thus he spoke and more, and inspiring confidence by swearing an oath, the crow made friendship with the mouse, who ventured out. The mouse brought pieces of meat and grains of rice in addition and there the two remained eating happily together.

Then one day the crow said to his friend the mouse: 'There is a river, my friend, running through the middle of the forest not far from here and in it there lives a friend of mine, a tortoise, Mantharaka by name; for his sake I shall go to that place where meat and food is easily obtained; here food is difficult to get and I am in constant fear of the hunters.'

In answer to the crow's proposition, the mouse said: 'Then let us live together! Take me there also, for I also have a dislike for this spot, but I will tell you about it when we get there.'

At these words Laghupātin took Hiranyaka up in his beak and went through the sky to the bank of the forest river. Meeting with his friend the tortoise Mantharaka, who afforded hospitality, he remained there together with the mouse. The crow then, among other stories, told the tortoise the reason for his coming and how it was connected with the story of his friendship for Hiranyaka. Then the tortoise made the mouse his friend in equal rank with the crow and inquired of him the reason for his disgust with his country home. Thereupon, with both the crow and the tortoise for an audience, Hiranyaka related the following tale concerning his own adventures:

'While I was living in a great hole close by to a city, there one night I brought a necklace from the king's palace and stored it up. As a result of looking at that necklace, which must have had magic qualities, I gained great strength, and the mice surrounded me as I now was well able to hunt for food. In course of time there came a certain religious ascetic who made his hut near our hole and lived by various alms of food. At night-time, after he had eaten, he would place what remained of his alms-food in his begging bowl on an inaccessible post with the idea of eating it the next morning. Then, while he was sleeping, I would enter through a hole and making a jump upwards I would bring the leftovers back every night. One time then, another wandering ascetic, a friend of his, came there and after they had eaten, they exchanged stories during the night. When I set out to bring back food, the mendicant, who was listening, made his bowl resound again and again with a piece of broken bamboo.

'"Why do you break off in order to do this?" the wandering ascetic asked, whereupon the other ascetic answered him: "An enemy of mine has appeared here, a mouse, and even though my food is sitting high up, he jumps up and takes it away from me and so I am frightening him by beating with the bamboo on my food bowl."

'"It is greed which really makes for the undoing of creatures," said the other ascetic in answer to his words. "Now listen
to a story which illustrates this. In a forest somewhere, a hunter,
after he had finished hunting, set up an automatic bow and
putting some flesh on the bow, he ran off after a wild boar.
Though he pierced the boar with an arrow, he himself was
struck and so severely wounded by the boar that he died. A
jackal was watching all this from a distance. He then came up
and, even though he was afflicted with hunger, he wanted to
hoard up the meat of the hunter and boar, and so he did not
eat any of it despite its abundance. He set off then and went to
eat what was lying on the bow and at that very instant he
was pierced by the arrow springing up from the machine and
died. So, you see, greed does not make for greater enjoyment
but only for misery." When he had said this the visiting ascetic
went on to relate: "If there is a spade here, give it to me so that
I may now duly prevent the mischief which you are being caused
by the mouse."

'When the resident hermit heard this he gave him a small
spade, and as I saw this from a concealed spot I entered my
hole. Then that rogue of a visiting ascetic, on seeing the hole of
my connecting passage, set in to dig away with the spade. He
kept on digging while I kept fleeing farther in, until at last he
reached the place where the necklace was, as well as the rest
of my stores. As I listened he said to the resident ascetic: "This
mouse has acquired such great strength through the magnificence of that necklace."

'And after they had taken all my possessions, and placed the
necklace by their heads, both the visiting and the resident
ascetics, being delighted with themselves, went to sleep. While
they were asleep, I came out again to recover it but the resident
ascetic woke up and caught me a blow on the head with his
stick. So, although wounded, I did not die, thanks to fate, but
entered my hole, though no longer did I have the strength to
jump up in order to get food. For the riches of men lie in their
youth and for want of it they grow old, and with it is lost also

spirit, strength, beauty and endeavour. So when my retinue of mice perceived that I was now intent on foraging only for myself, they left me and all went off. So do servants leave a master without means, bees a tree without flowers, swans a lake without water though for a long time they have lived there.

'In this way, being in distress for a long time there, I gained Laghupātin here for a friend, O noblest of tortoises, and so have come to your presence.'

When Hiranyaka had told this, the tortoise Mantharaka declared: 'Look upon this as your own home, so do not, my friend, venture restlessly again. There is nowhere a strange land to the worthy man, nor discontent for him who is content, nor misfortune for him who is resolute, nor yet again anything unattainable to him who tries.'

While the tortoise was saying this, a deer called Chitrānga came from a distant spot to that forest, having been terrified by hunters. When the tortoise and the others saw him and realized that no longer was any hunter coming after him, they comforted him and made friends with him. Then they all lived there as friends, the crow, the tortoise, the deer and the mouse and remained happy in helping one another out.

One day, however, Laghupātin, in order to find why Chitrānga was late coming back, perched on a tree and looked out over the forest. Then he saw Chitrānga tied up in a deadly snare on the bank of a river, so, coming down again, he told the tortoise and the mouse. Then after taking counsel together, Laghupātin took the mouse Hiranyaka up in his beak and brought him to the place where Chitrānga was. And the mouse Hiranyaka straightway set in to free the deer who was miserable at being tied up, comforting him by chewing away at the bonds and cutting them. In the meantime the tortoise Mantharaka had come up along through the river out of affection for his friends and climbed up on the bank near where they were. At that moment the hunter who had laid the snare came up unexpectedly, but while the deer and the others ran off, he caught and captured the tortoise by throwing his net around him. But just as the hunter was going away, distressed at the loss of the deer, the mouse, who with his long sight had observed all this, commanded the deer to go a little way off and fall down pretending to be dead. Meanwhile the crow perched on his head pretending to peck at his eyes. When the hunter saw this he considered the deer to be dead and so in the bag, whereupon he set off towards it, leaving the tortoise down on the river bank. When the mouse saw him going he went up and bit through the net round the tortoise, who, now that he had been freed by him, slipped into the river. The deer, for his part, when he saw the hunter coming near without the tortoise, got up and fled, while the crow flew up into a tree. When the hunter turned back then, he saw that the tortoise had escaped from the chewed up net, and he went home lamenting, knowing well that an animal which has fled through fear cannot be caught.

Then the tortoise and the others gathered together there in great delight, and the deer out of gratitude said to the tortoise and the other two: 'I am indeed fortunate in gaining you as friends, for today, without regard for your own lives, you have rescued me from death.'

After these words of praise, the deer continued to dwell happily with the crow, the tortoise and the mouse, and they all enjoyed their friendship for one another.

## The Mongoose, the Owl, the Cat and the Mouse

THERE was once upon a time a mighty *banyan* tree on the outskirts of the city of Vidishā, and four creatures lived, each with its own abode, in that great tree: a mongoose, an owl, a cat and a mouse. The mongoose and the mouse dwelt in different holes in the root of the tree, while the cat lived in a great hollow half-way up the tree and the owl was in a nest of creepers at the very top, unapproachable by the others. The mouse was liable to be killed by all three whereas the cat was able to kill the other three. So out of fear of the cat, the mouse and the mongoose, as well as the owl on account of his very nature, wandered around at night, all three of them seeking food. The cat, however, wandered about in a barley field near by both by day and night, fearless, ever on the watch to catch the mouse; the others, however, went discreetly at certain times in their quest for food.

One day there came a certain hunter to that place and when he saw the tracks made by the cat going into the field, he set snares round the field in order to catch it and then went away. When the cat came out in the night then with the intention of killing the mouse, he stepped into one of the snares and was caught in it. Then the mouse came to the spot while he was secretly searching for food and when he saw the cat caught he was delighted and danced for joy. Then, as he was entering the field, the owl and the mongoose came from afar along the same track and when they saw the cat caught they resolved to seize the mouse. The mouse, however, when he saw these two coming in the distance was terrified and thought to himself:

If I take refuge with the cat who strikes fear into the owl and the mongoose, then even though he is tied up, my foe will kill me with a single blow; and if I go any distance from the cat then the owl and the mongoose will slay me, so beset around with enemies, where will I go, what shall I do? Well, I will take refuge with the cat, for he has got into difficulties and he may protect me for the sake of saving himself if I undertake to nibble through his bonds.

With these thoughts in his mind, the mouse gently approached the cat and said to him:

'I am very sorry indeed that you have been caught, so I shall gnaw through your bonds, for on account of being neighbours worthy people love even their enemies, yet I have no confidence in you since I do not know your thoughts.'

When he heard this the cat said: 'From today onwards you shall be my friend since you have given me my life.'

At these words from the cat, he took shelter near his body, so that when the owl and the mongoose saw what had happened they lost hope and went away.

The cat then being troubled by the snare, said to the mouse:

'The night has mostly gone, my friend, so quickly cut my bonds.'

The mouse for his part went on slowly nibbling away, keeping an eye out for the hunter's approach, and pretending to make a long business of it by continually chewing away with his teeth. Suddenly as night turned into dawn, the hunter came along and as the cat pleaded, the mouse quickly cut through the thongs. As soon as his bonds were cut the cat ran away in terror from the hunter, and the mouse, being free from the threat of death, fled and entered his hole. He had no confidence either when he was afterwards called upon by the cat, for he announced:

'On a particular occasion an enemy becomes a friend by chance, but it does not last for ever.'

# The Earthquake

ONCE upon a time there was a palm grove studded with *bel* trees lying close to the Western Sea, and there a hare lived under a small palm shrub at the foot of a *bel* tree. One day, after bringing his food back to the foot of the palm shrub, the hare thought to himself:

If the earth were to fall to pieces, what would become of me?

At that very moment a huge ripe *bel* fruit fell right on top of the palm shrub. With this the little animal jumped straight up in the air, being quite convinced that the earth was indeed falling to pieces around him, and fled madly away without ever looking behind him. Another hare who saw him flying along scared to death asked him:

'Why are you running away looking so terrified?'

'Don't ask me.'

'But why, but why?' the other repeated, racing after him. The first hare, without looking back, said:

'The earth is falling to pieces behind us.'

So the other fled after him. In the same way another saw him and yet another until a hundred thousand hares were all flying in a bunch after him. Then a deer saw them, and a boar, and an antelope, and a buffalo, and a gayal, and a rhinoceros, and a tiger, and a lion, and an elephant and all asked:

'What is this?'

'The earth is falling to pieces!' was the reply, so they all joined in the stampede. In this way gradually the line of animals stretched out over a league.

Then a great lion appeared and he saw that wild flight of animals and asked:

'What is this?'

When he was told that the earth was falling to pieces he thought:

There is no question of an earthquake, but I suppose they must have misunderstood some sound they have heard and if I do not make some great effort, they will all come to destruction. I must save their lives.

With this he sprang with his lion's speed to the foot of a hill in front of them and three times roared his lion's roar. They were all frightened out of their wits at the sound of the lion, so they turned about, huddled up together and stayed where they were. The lion padded in between their ranks and asked them:

'Why are you fleeing?'

'The earth is falling to pieces.'

'Who has seen it falling to pieces?'

'The elephants know about it.'

He asked the elephants, but they said: 'We know nothing about it, it is the lions who know.'

But the lions said: 'We know nothing about it, it is the tigers who know.'

The tigers said: 'The rhinoceroses know.'

The rhinoceroses said: 'The gayals.'

The gayals: 'The buffaloes.'

The buffaloes: 'The antelopes.'

The antelopes: 'The boars.'

The boars: 'The deer.'

But the deer said: 'We do not know, ask the hares.'

When the hares were asked, they all pointed to the one hare and said:

'This one told us.'

So the lion asked him: 'Is it really true that the earth is falling to pieces?'

'But most certainly it is true; I saw it.'

'Where were you living when you saw this?' the lion asked.

'Oh sir, in a palm grove studded with *bel* trees, close to the Western Sea. There while resting under a palm leaf in a palm shrub at the foot of a *bel* tree I began to wonder where I would go if ever the earth should fall to pieces and lo! at that very moment I heard the crash of an earthquake and so I fled for all I was worth.'

The lion thought to himself: No doubt a ripe *bel* fruit fell from above on to the palm leaf and made a crash, so that this little fellow on hearing it thought that the earth was cracking up and he fled. I had better look into the matter.

He thereupon took the hare aside and spoke words of re-assurance to the great herd of animals, saying:

'I shall go to find the exact truth about this earthquake, whether in fact such a thing did or did not happen where he says it did, and you are all to stay here until I come back.'

With that he put the hare up on his back, galloped to the palm grove with his lion's speed and put the hare down.

'Go now,' he said. 'Show the place you were talking about.'

'Oh sir, I would not dare.'

'Come along, don't be afraid!'

The hare, not daring to go near the *bel* tree, stood some way off and, pointing out to the lion the spot where he had heard the terrible crash, he said:

> 'There was a crash! May you fare well
> On seeing the place in which I dwell.
> But I for certain cannot say
> Where the cause of trouble lay.'

At these words the lion stepped up to the root of the *bel* tree and seeing that the spot under the palm leaf where the hare had been was just exactly where a ripe *bel* fruit had fallen on top of the palm shrub, he then carefully made sure that there was no sign of an earthquake. After this he put the hare on his back and, returning with lion-like speed in all haste to the assembly of animals, he told them all that had happened.

'You need have no fear now.' Thus he reassured them and dismissed them all. Indeed if it had not been for the lion on that occasion, the animals would all have fallen into the sea and would have drowned, but thanks to the lion they were all saved.

## The Hawk's Friends

WHEN King Brahmadatta was reigning in Benares, certain bands of roving thieves on the borders of the country were in the habit of setting up their quarters wherever they could find abundance of venison, and they brought up their families by hunting in the forest, killing the deer and other game therein and eating the meat. Not far from where they had settled was a vast natural lake; on its southern bank there dwelt a hawk, on the western bank a female hawk, on the northern bank a lion, the king of beasts and on the eastern shore an osprey, while on an island in the centre of the lake a tortoise had made his home.

The hawk one day asked the female hawk to be his wife, but she questioned him first as to whether he had any friends.

'My dear, I am afraid I have not,' he answered.

'One should have friends to help one in case of any danger or misfortune which might arise,' she said to him, 'so go and get yourself some friends.'

'With whom shall I make friends?'

'Make friends with the osprey who lives on the eastern shore and with the lion who lives on the northern shore and with the tortoise who lives in the middle of the lake.'

He agreed with what she had said and accordingly did so. Then the two of them set up house together in a *chadamba* tree which grew on a little island in the lake and, making their nest there, they dwelt happily. In course of time two baby hawks appeared and while they were still in the nest, along came the borderers who had been out hunting a whole day in the forest and had caught nothing.

'We cannot go home empty-handed,' they were saying amongst themselves, 'so let us try to catch some fish or tortoises.' They therefore set out on the lake and landed on the little island, sitting down at the root of the *chadamba* tree. There they were eaten by all kinds of insects and in order to drive them away they rubbed fire sticks together, made a fire and sent up clouds of smoke. As the smoke billowed up, it surrounded the birds, and the young birds began to cry. When the borderers heard the noise they said:

'What luck! There is the sound of some young birds. Let us get up and take firebrands, for we are so hungry we cannot sleep, but if we get something to eat, then we will be able to sleep.'

With this they stirred up the fire and seized burning firebrands. When the hen bird heard what they were about and realized that the men would eat her children, she knew that the time had come to ask her friends to rescue them from the danger, so she decided to send her husband to the osprey.

'Go, my lord,' she said, 'tell the osprey that a danger threatens our children. Tell him this: these wild men seize firebrands on the island; they want to eat my children. Tell the tale, O hawk, to friends and allies; tell the birds of the destruction of our kin.'

The hawk flew swiftly to where the osprey lived and, uttering his call, made his arrival known. Being invited, he approached and made his salutation. When he was asked why he had come, he spoke as follows:

'You are, O bird, the most excellent of all the beings that fly; I come, O king osprey, to you for help; savage hunters want to eat my children; please be a source of comfort to me.'

The noble osprey comforted the hawk, saying to him: 'Have no fear!' and then he spoke as follows:

'In good times and in bad, wise men who seek well-being make friends and allies; I will do something to help you, O hawk; for the noble perform duties for the noble.'

Then he asked him: 'Have the rogues climbed the tree yet?'

'No, they have not climbed it yet, they are just brandishing torches.'

'Then go quickly, comfort my good friend your wife and tell her to expect my arrival.'

He did so, whereupon the osprey departed forthwith and, alighting on a *chadamba* tree not far off, he watched the rogues climbing up. Then when he saw one of the borderers climb to nearly within reach of the nest, he plunged into the lake and, filling his beak and wings with water, he poured it down on to the firebrand so that it was extinguished. The rogues however, determined to eat the hawk and her young, climbed down again and, lighting firebrands anew, started up the tree again. Once more the osprey put the brands out in the same way, and, as he kept on doing this, night came on. He became very much exhausted, and the feathers of his breast were singed and his eyes were red with the smoke.

When the hawk saw him, she said to her husband: 'The osprey is completely worn out; go and tell the tortoise so that he may get some rest.'

When he had heard what she had to say, he approached the osprey and spoke to him as follows:

'Whatever duties he who is compassionate and noble has to perform for the noble, that you have done; look after yourself now, do not burn yourself; with you still alive we shall save our children.'

When he heard this, the osprey spoke the following words in a voice as loud as a lion:

'Even though wounded in body, I am not afraid to stand this guard for you; thus they act, friends for friends: they abandon their lives; such is the law of the good.'

The hawk begged him, however, to take a little rest and, flying off to where the tortoise lived, he woke him up.

'Why have you come?' the tortoise asked him and was told about the terrible danger and how the osprey had been doing all he could about it ever since the morning, but that now he was getting weary.

'That is why I have come,' the hawk went on, and then he spoke as follows:

'Some, though fallen through their own misdeeds, rise again through the compassion of friends; my children are threatened and that is why I have come to you; you who live in the water, please do a favour for me.'

When he heard this the tortoise replied as follows:

'The wise make friends and allies by means of wealth or food or simply through their own true nature; this favour I will do for you, O hawk, for the noble perform duties for the noble.'

Then the son of the tortoise who was standing close by heard what his father had said and thought that his father should not be troubled over this matter and that he himself should perform the deed, so he next spoke up and said:

'You stay here untroubled, O my father; the son should carry out tasks for the father; I shall do this for you and protect the children of the hawk.'

But his father replied with the following words:

'Truly indeed this is the custom of the good that a son carries out duties for his father; yet when they have but seen me, fully grown as I am, they will not plague the children of the hawk.'

When he had said this the great tortoise sent the hawk off, telling him not to fear and that he would follow him. Thereupon he dived into the water and collected up some mud which he took with him to the little island and threw on the fire.

When the rogues saw him lying there, they said to themselves: 'Why need we bother about the young hawks? Let us just turn this old tortoise over and kill it and it will suffice for us all.'

They gathered creepers and got hold of some thongs, but

even when they had bound the tortoise in several places by dint of strips from the clothes they were wearing, they still were unable to turn him over on his back. The tortoise then went off, dragging them with him, and plunged into some deep water. They were so eager to catch him that they fell into the water after him and being exhausted from swallowing huge quantities of water, they crawled back to land again.

'Look what has happened,' one said, 'for half the night an osprey has kept putting out the fire and now a tortoise has dragged us into the water and made us swallow water until our stomachs nearly burst, so let us light the fire again and at dawn we can eat the young hawks.'

Thereupon they began to make a fire. When the female hawk heard what they were saying, she turned to her husband:

'My husband, sooner or later these men will come and eat our young ones, please go therefore to our friend the lion.'

He flew off that very instant and when he approached the lion he was asked by him why he had come at such a strange hour. He told the lion about everything that had happened and then said he:

'O mightiest of beasts and men, both man and beast run to the strongest when beset by danger; my children are threatened, that is why I have come; you are our king, so be a source of comfort to me.'

When he heard that, the lion said:

'I shall do this for you, O hawk; let us go and slay these enemies of yours; for how could anyone wise and considerate fail to stir himself in order to protect a friend?'

When he had said this he urged the hawk to go and comfort his children and then he set out himself, scattering the crystal water. As the rogues saw him coming, they cried out:

'Our fire has been put out by an osprey; the clothes we were wearing have been taken by a tortoise, and now that we have nothing left a lion turns up to threaten our very lives.'

Scared to death they ran off in all directions. When the lion arrived at the foot of the tree he did not see a soul, except for the osprey, the tortoise and the hawk who came up to greet him. He spoke to them about the blessings of friendship and advised them with these words:

'From this time forward take heed that you never break the ties of friendship.' With this he left them and they each returned to their own homes. As the hen hawk looked at her young ones, she realized that they had been kept safe through friends and she and her husband talked long and happily about it.

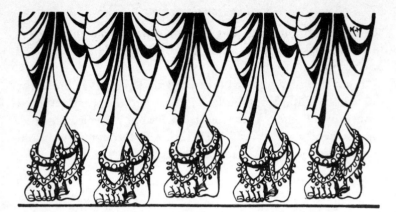

## Nala and Damayantī

I N the land of Nishadha there was a young king named Nala, the son of Vīrasena. He was deft in subduing the wildest stallion and supreme among the archers in his realm. Nala knew the sacred lore and was leader of a mighty host; honouring truth and justice, he held sway over the chiefs of his land, as the sun the sky. Admired by many a fair princess, his beauty so outshone his fellow mortals that even Kāma, god of love, envied him his human form. Yet Nala remained unwed.

At that time too there was a ruler in Vidarbha, a king named Bhīma, whose ferocious power brought mighty vassals tumbling to obey his behest. Bhīma and his queen were blessed with three noble sons, and yet a fourth child, a daughter, Damayantī by name. When Damayantī grew of age, the fame of her beauty spread like the scent of jasmine throughout the kingdom and beyond. Of fair complexion and slender waist, she rivalled the long-eyed Shrī, goddess of beauty, while, in her palace, surrounded by a hundred female slaves, the radiance of her charms so held the minds of those who saw her that the moon itself trailed, forgotten, in the sky.

So it happened that Nala's envoys returned home glowing with tales of Damayantī's loveliness, while in Bhīma's palace, the maidens praised Nala endlessly in Damayantī's presence. There soon arose between these two a love that grew and bound them both, though neither yet had seen the other. One day while Nala, silent in his love, wandered in the gardens of his palace, a flock of royal swans, their plumes all touched with gold, strutted before his gaze. He caught one, and thereupon the bird spoke to him with human voice:

'You must not slay me, mighty Nala, for I will so speak to Damayantī that she will never think of any man other than yourself.'

As Nala released the bird, the whole flock wheeled into the sky and set course straightway for Vidarbha.

When they reached Vidarbha's capital, the royal swans alighted in Damayantī's garden, drawing wonder from the princess and her maidens. As the girls began in play to chase the golden birds, the swan that Damayantī followed addressed these words to her:

'In Nishadha, fair maiden, there dwells King Nala, matchless among mortals; he it is whom you should marry. For we have seen those who dwell in the three worlds, above, upon and beneath the earth, yet such worth as his we have not seen. It is fitting that he, a peerless hero, should wed a peerless bride.'

When Damayantī heard this, she begged the golden bird to speak on her behalf to Nala, whereupon the noble swan retraced his lofty path to Nishadha and related all to Nala.

From that day on, her maidens saw a change in Damayantī. Once carefree, she now spent days deep in thought; her bright face became sad, and no longer did she find refuge in friendships, in banquets or in sleep. The report was swiftly brought to Bhīma, who, reflecting that his daughter was now of an age to marry, determined instantly to hold a *svayamvara* festival, a festival where she might choose a husband for herself. Bhīma spoke, 'Heroes, let this *svayamvara* be well attended!' Thereupon the earth resounded with the noise of royal retinues, of

elephants, of horses and of chariots, as all the mighty kings, resplendent with their armies, drew in towards Bhīma's city; and
that generous monarch received each and every suitor with all
the honours that were fitting for such guests.

At that same time, the world of Indra, chief amongst the
gods, was visited by two most noble seers, Nārada and Parvata,
who, their age-long penance done, received a royal welcome.
In answer to Indra's courteous greetings, Nārada replied:

'We fare well, immortal lord, and the kings on earth fare
well.'

Whereupon the slayer of countless demons instantly re-
torted:

'How is it then that I do not see these noble warriors approach the gates of heaven? They whose destiny it is to meet
death with unaverted eye, why do I not see them, those dear
and worthy guests?'

'There is a fair princess,' Nārada replied, 'Damayantī, daughter of King Bhīma, a more lovely maiden none on earth. All the
kings and princes, setting aside their strife against evil, flock
to her *svayamvara*, each one anxious in the hope to be her
choice.'

At these words all the gods drew near and with one en-
chanted mind they spoke, 'Let us go there as well!' Yet while
they sped their glorious chariots headlong from the surface of
the sky, those immortals saw, far below on earth, a being more
of the world they left than any mortal being. The steeds were
checked, the chariots hung in mid-air, while the gods looked
silently at Nala, for it was he, upon the road to Bhīma's royal
city. Then drifting downwards gently through the air, they
said to the king as he sped towards his beloved:

'Royal Nala, famed for your honesty, we entrust you with a
message. Bear it faithfully for us!'

Nala, with his hands folded in respectful greeting, spoke:

'I promise to do your bidding, but tell me whose messenger
I am, and what is the message I must bear.'

'Know us to be the immortals,' Indra replied, 'for I am Indra, and this is Agni, god of fire; here is Varuna, god of waters, and here Yama, he who sits in judgement over those about to die. We bid you go to Damayantī to deliver her this message: the guardians of the world, Indra, Agni, Varuna and Yama come to your *svayamvara*; choose one of them as husband!'

Nala, still with folded hands, replied:

'I who journey for the selfsame purpose, how can I fulfil your aim? Spare me from this task, most mighty lords!'

But the gods reminded Nala, '"I promise to do your bidding" were the very words you spoke; how can you now not do it? Go at once!'

Yet Nala still was hesitant. 'How shall I enter Bhīma's fully guarded palace?' he asked.

'You shall enter,' Indra spoke and sent Nala upon his way.

Near the close of day, Damayantī's maidens were sitting in the court when, of a sudden, wonderstruck, they saw a god-like prince appearing in their midst. Rising from their seats, being unable in their embarrassment to speak a single word, they wondered all the while who he might be and, silent at so much beauty in one man, they worshipped him with their enkindled hearts. A bright flame darting from the fire of love, Damayantī, sweetly smiling, came towards Nala and, as she smiled at him, his love, although constrained within his heart by his duty to the gods, increased still more, and he, too, smiled at her. Then Damayantī asked him in her wonderment:

'Who are you, most handsome prince, that you exalt what lies within my heart? I wish to know how you, a blameless hero, have entered here as might a god. How did you come and why were you not perceived? For my father, the king, terrible in his command, has had my dwelling fully guarded.'

'Know me to be Nala, most lovely maiden,' Nala replied, 'a messenger from the gods. The gods Indra, Agni, Varuna and Yama desire to win you; choose one of them as husband, fair princess. Through magic power given to me by the immortals, I entered unperceived. No one saw me nor barred the way. Now

that you have heard my tidings, decide, fair maiden, according
to your wish.'

Paying reverence to the unseen gods with folded hands, she
laughed and said:

'You, although a king, should speak out in fullest trust to me,
so that I may help you. Both I and all that I possess belong entirely to you, so place your thoughts in my care, my lord. Those
words I heard from the swans, my prince, still burn me. Only
for the sake of you, my hero, were all the kings assembled here.
If you so belie your compassionate nature that you refuse me,
then I will resort to poison, fire, water and the noose; all this
because of you.'

'While these mighty guardians of the world await your answer,' Nala replied, 'how can you want a mere human? I am
not even as the dust on the feet of those creators of the world, so
turn your mind on them. Death is the lot of the mortal who in-
curs the wrath of heaven. Save me, O most faultless maiden,
and choose those powerful gods. You will then enjoy clothes
free from dust, bright garlands of unfading flowers and the finest
jewels.'

Damayantī, with tears welling up from this hurt, replied:
'To all those gods I have paid due reverence, but you, lord of
the earth, I choose as husband; this is my plighted word.'

The king then, trembling, with his hands respectfully folded,
said: 'As I have made this special promise on the gods' behalf,
I must carry out my duty. If then the chance arises, I shall
plead on my own behalf. Then, fair maiden, it is for you to
decide.'

Whereupon a sweet smile appeared on Damayantī's face, and
slowly, while the tears still choked her words, she answered Nala
the king: 'I have seen a way open to us, O lord of men, whereby
no blame will fall upon my king. Let all of you, you yourself
together with the gods, come to my *svayamvara*. I shall then
choose you, my powerful lord of men, in the presence of those
protectors of the world. In this way you will not be at fault.'

Thereupon Nala returned to the gods who were eagerly awaiting his news.

'Did you see her?' they asked. 'What did Damayantī say, with her sweet smile? Tell us everything that happened.'

Nala related all that had taken place; how Damayantī had chosen him despite his faithful execution of the message, and what her final words had been. '"Let all the gods come to my *svayamvara*, together with yourself, my powerful lord, and in their presence I shall choose you, great Nala. In this way no fault will be yours." These were her words,' said Nala. 'As for what is to follow, you, the chiefs of the gods, must determine your verdict.'

The bright day came when the constellations took up auspicious places in the heavens, and Bhīma summoned the kings. Their wonted swiftness in obeying the mighty ruler could scarce disguise the eager haste of those love-tormented monarchs who filed, like lions under the forest trees, through the golden archways into the jewelled hall. There they sat, each upon a throne, all bedecked with fragrant garlands and making the air like the night sky with their gem-encrusted ornaments.

Then Damayantī entered the assembly, and her lovely face stole the eyes and the minds of all those kings. A thousand glances fell upon that faultless form and there they stayed and did not waver.

While each king was being announced by name, Damayantī of a sudden saw five men standing there, all alike in form. Look as she might, there was nothing that could distinguish them, for, as she stared at each, it was Nala whom she saw.

How shall I tell which is really Nala, she wondered, and how shall I recognize the four gods who have taken his form? But while she recollected what she had heard from wise men about the signs that distinguish a god, she knew her task was hopeless, for no such signs were to be seen on any of the five.

It was now the time, she thought, to put her trust in the

gods, and so, paying the deepest homage with voice and mind, she folded her hands and, trembling, uttered a prayer:

> 'It was when I heard
> > the swan's word
> That I chose Nala as my betrothed.
> Pray, O gods, reveal him to me.

> 'Nor sin have I wrought
> > by word or thought
> When I chose Nala as my betrothed.
> Pray, O gods, reveal him to me.

> 'O gods from above
> > you thought this love
> That gave me Nala as my betrothed.
> Pray, O gods, reveal him to me.

> 'For I took the vows
> > that love allows
> When I chose Nala as my betrothed.
> Pray, O gods, reveal him to me.

> 'A cast of his clay
> > allows no way
> To know my Nala, my own betrothed.
> Pray, O gods, show your own true selves.'

When they heard that sad prayer, the gods did all that lay within their power to reveal themselves. Damayantī then saw all those divine intelligences, for on their bodies, no sweat came; the flowers in their garlands were as if fresh-picked; their eyes did not move nor their feet touch the ground. But Nala was discovered easily, for his feet touched the ground, his eyes moved, and his body, flecked with dust and sweat, cast a dark shadow. The long-eyed maiden went swiftly to his side and, holding the hem of his robe, she cast a garland of the fairest flowers about his shoulders. Thus Damayantī chose her lord.

The kings sorrowed, but the gods and the wise rejoiced.

Then King Nala, son of Vīrasena, in his great delight, spoke words of comfort to his bride:

'Since you with your sweet smile have chosen me before the gods, I shall belong to you for as long as there is life in my body; this I swear to you.'

Thereupon the pair, happy in their love, turned towards the gods; and the gods, being pleased, bestowed on Nala eightfold blessing. Indra granted him the power to see the divine presence whenever he performed religious sacrifices, as well as the gift of treading all paths most uprightly; Agni gave him fire to command at will, adding the gift of unharmed entry to his bright abodes. Yama bestowed a skilled taste in food and the right to the highest place, while Varuna gave him water to command at will and garlands rich in faultless fragrance. With this, the gods returned to heaven, and the kings, recovered from their loss, returned to their own lands.

King Bhīma arranged the wedding with all due rites, and Nala and Damayantī lived for a while in his city after their marriage. Then Nala took leave of his father-in-law and journeyed with his wife back to the land of Nishadha. There they settled, enjoying every happiness; and soon were blessed with a son, Indrasena, and a daughter, Indrasenā.

Yet as the gods were returning to their mansions in the sky, they saw an evil pair approach, and Indra, chief of the gods, addressed the senior of the two: 'O Kali, where are you going, with your ally Dvāpara?'

'We go to Damayantī's *svayamvara*, for I intend to make her my bride.'

Laughingly Indra answered: 'Too late you come; the *svayamvara* is over; and she has chosen Nala from our side.'

Then the demon Kali in the wildness of his rage swore that he would bring foul punishment on Nala and his happiness. But the gods warned him that his cursing was in vain and they went upon their way.

Kali turned then to his ally and said: 'I vow that I will cast

Nala from his kingdom, from his happiness with his bride. You, Dvāpara, must lurk within the dice, and that is how we shall win our aim.'

For twelve long years the two scoundrels haunted Nala's palace until at last one day they saw their chance. Nala went for prayers with feet unwashed and straightway Kali seized that very moment to enter into Nala's inmost heart. He then called Pushkara, Nala's brother. 'Come, Pushkara,' said he, 'play with Nala; by my aid you shall win his entire kingdom.'

So Pushkara approached Nala, saying: 'Let us play with the dice, my brother.' Nala, ever fond of gaming, could not resist the challenge; all he staked upon the throws, his riches, his horses, and all he lost. The citizens came to the gates, the counsellors assembled, the charioteer was sent as spokesman to Damayantī to beg her make her lord desist. Yet all remained in vain; not even the sombre dew of his fair queen's tears kept Nala from the game. Fair Damayantī returned again to Vārshneya, the charioteer, and persuaded him, 'My lord will no longer listen to what I say; so do you take the two children back to my father's city, back to Bhīma's city.' While the good Vārshneya drove with Indrasena and Indrasenā to the kingdom of Vidarbha the dice were still being thrown, and over many a terrible month the treasures, the palace, the kingdom, all of Nishadha was lost.

When Pushkara gained the kingdom, he instantly decreed that anyone befriending Nala should be put to cruellest death. Thus Nala wandered forth, outside the city gates, clad in but a single garment, all his fine raiment gone. Damayantī followed him slowly and saw him try to capture some golden-winged birds. Nala cast his garment over them, but they, mocking, flew up and tore the cloth away. 'O king,' they said, 'we are the dice and we have come to take your last possession.'

When Nala saw that not a single thing remained to him, he turned to his sad wife and said: 'Now that I have nothing, you must return to your father's kingdom, for I no longer can look after you.'

Yet plead as he might, his wife would not leave his side nor cease to entreat him to return with her to her father's kingdom. Wearied out by walking, thirst and hunger, they drew near to a humble cabin in the woods and there, in the mire, for there was no mat to rest on, they laid themselves down to rest, covered only by one garment. Yet as Damayantī fell into deep slumber, the dread thought of all that he had lost kept sleep from comforting Nala. Round and round the cabin he paced, torn between the thought of staying by his wife's side, or else abandoning her and hoping she might reach her father's realm. As he paced to and fro, suddenly he espied a shining sword hanging from a beam and with it he cleft their garment in half. Then placing half over his sleeping queen and drawing the remaining half about himself, he made to flee from the cabin, but ever and anon he returned. In the end, possessed completely by Kali, he sped away.

> Whom sun nor wind knew not,
> Now on the earth lies bare;
> Whose tender smile sleep shrouds,
> What fate will fall to her share?

> Her beauty left by me
> Ill in the wood will fare.
> O sun and winds of heaven,
> I have left her heart in your care!

A short while after Nala had gone, Damayantī awoke and, finding her lord gone, she shrieked in anguish. Turning this way and that, thinking to see her lord hidden behind this tree and that, the fair princess ran frantically to and fro. And as she wandered she cursed the evil spirit that had possessed her lord and driven him from his kingdom. Suddenly, as she passed by the crevices in some rocks, a gigantic serpent crept towards her and wrapped her in its coils. But her only words were for her lord, and who would stand by him when once his senses returned. As she thus lamented, a forest huntsman heard her cries and, running swiftly towards her, dispatched the serpent

with a mortal shaft. Smitten by her beauty, alone and un-
protected, he would have carried her away, but Damayantī,
knowing his false intent, straightway let fall her curse on him
for trying to make her fail her lord. On the instant the hunter
fell lifeless at her feet. Then she wandered, the daughter of
Bhīma, three whole days and three whole nights, ever and anon
calling to her lord, hoping over and over again to hear his con-
soling voice. The very mountain lions stood respectfully aside
from her path, owning in her fair form the sovereignty of Nala
the king, and turned down towards the river, while she climbed
slowly up towards Vīrasena's mountain, where her dear lord's
father lay buried. Yet three days more and three nights, ever
searching and ever calling, she came suddenly upon three her-
mits in their cells. And when they asked her who she was, she
told them the whole sad tale. No sooner had they heard it than
with one voice they prophesied: 'The time will come, fair lady,
when Nala once again will rule justly in his city, with you devoted at his side.' And with these words, they vanished, her-
mits, sacred fires and cells, leaving Damayantī long in wonderment as she wandered on and on, past *ashoka* trees, hills and
many streams. Then, by one vast and peaceful river she came
across a caravan, with merchants, horses, elephants and wagons. At the sight of her, scant-clad and dishevelled, some took
her for a witch and fled in terror, but stout Shuchi, leader of
the host, held his ground and answered her that nowhere had
they seen Nala in that wild and trackless region, but that if she
liked, she might join them, bound as they were for the kingdom
of Subāhu, king of the Chedis. So on they journeyed and at
nightfall reached a beautiful lake, where, after feeding their
animals and then themselves, the whole company lay down to
sleep. Alas, however, in the very dead of night a herd of wild
elephant came by and, as soon as they scented the tame beasts
of the caravan, they tore this way and that through the encamp-
ment, trampling and crushing many of the merchants. Loud
cries of agony filled the night air, even curses against the luckless woman upon whom all blame was put. As soon as Dama-

yantī heard herself thus reviled she fled in terror from the camp, praying as she escaped for some sign from the gods to explain the hard fate she had brought upon her protectors. But no sign came, nor did she meet anyone save some Brahmans who kept her company till they reached Subāhu's city. There Subāhu's mother saw her as she entered through the city gate, a lost and luckless figure yet with a splendid majesty of beauty. Straightway she sent her nurse down from the palace to lead the wanderer in. Once more Damayantī told the tale of her misfortunes, whereupon Subāhu's noble mother instantly promised her a safe home, the friendship of her daughter Sunandā and that men-at-arms would be sent forth to seek her husband.

Nala had not long left Damayantī before he beheld a huge fire blazing in the forest and heard a voice coming from the flames.

'Come quickly here, Nala, come quickly,' it said.

Nala rushed to the centre of the flames and saw the king of serpents lying coiled there.

'Know me to be Karkotaka, cursed to remain here till Nala should rescue me. Lift me up!' As the serpent spoke he shrank to a mere finger's size. Nala lifted him and brought him clear of the flames when once again the serpent spoke to him: 'Go forward now, counting your steps, and I shall give you all good fortune.'

As Nala counted the tenth step, the serpent bit him and all at once his regal form was changed.

'Take comfort,' spoke the serpent, 'for Kali who afflicts your body will dwell in torment within you now. Go forth without fear, O king! Be known henceforth as "Vāhuka the charioteer"; go to Rituparna who rules in royal Ayodhyā. He will give you skill in dice and you will give him knowledge in taming steeds. In this way you will regain your kingdom, your wife, your children, and when you have achieved this, then remember me and don this garment which I now give you. So you will regain your own true form.' With that the serpent vanished.

Nala then set forth directly as the serpent had commanded him, and on the tenth day he entered Rituparna's city. Straightway he offered his services to the king:

'Vāhuka I, skilled in taming horses above all and skilled as

well in preparing food, these and many other things I offer to your service.'

Rituparna engaged him without further ado as chief of horse, giving him Vārshneya and Jīvala as aides. There Nala remained, engaged in his tasks, and every evening he uttered this single verse:

> 'Torn by hunger and by thirst,
>       Where rests my love's sweet face?
> Mindful of her foolish man,
>       Whose presence does she grace?'

Nor could Jīvala draw more from him but that he, through none but his own fault, had lost a peerless bride.

King Bhīma by now had sent forth Brahmans to every quarter to seek out King Nala and Damayanti his child, promising great rewards of a thousand kine and a grant for maintenance of a village should they be found, and even for news of them still a thousand kine. Not long time passed before a Brahman named Sudeva came to Chedipur and there he saw, standing beside Sunandā, his royal master's daughter, her beauty shining but wanly through the marks of her adventures. When good Sudeva had gazed upon her carefully, he knew it behoved him to speak words of comfort to her, to assure her that her parents and her children and her brethren, all were well. As soon as Damayanti knew him, she asked him repeatedly about all who were his friends. When Sunandā's mother observed this sincere and earnest discourse, she came at once to where Sudeva stood and spoke to him:

'Who is this lady, O Brahman? She appears known to thee. Whose wife is she? Please tell us all her adventurous tale.'

Sudeva began thereupon to tell her of Damayanti and Nala, whereupon the queen mother realized that Damayanti was her niece, since she herself and Damayanti's mother were sisters. Then Damayanti asked the queen mother if she might take her leave and journey back to where her children were. Thereupon she and Sudeva were supplied with a mighty retinue that they might travel safely back to Bhīma's realm. When they had returned, and Sudeva received his reward, Damayanti passed a night of rest and then pleaded with her mother:

'Let our chief task now be to bring my husband Nala home.'

Whereupon the queen persuaded Bhīma to send forth a further troop, yet ere they set out they gathered before Damayanti, who gave them this message:

'Wherever you find people assembled, may you repeat this speech, again and again:

"Where did you go then, you gambler,
     And cut half my garment to keep?
Why did you leave me, my loved one,
     Your love in the forest asleep?

"At your command she awaits you,
     She stays there and waits for your words,
Wasted away in her sorrow,
     A girl whom a half-garment girds.

"Silent she sighs in deep sadness,
     Tear-stricken for you, her dear lord;
Hero, please grant her a favour,
     And to her some small answer afford!

"Always supported and guarded
     Should be by her husband a wife;
Why did you fail both these duties,
     To whom all the law was your life?

"You who as merciful, noble
     And wise have for ever been famed,
Now have become without mercy,
     I fear, since my fortune is lamed.

"Tiger-man, grant me your comfort,
     O give me your pity, my lord,
Mercy is highest among duties,
     From you have I known this true word."

If anyone replies to this, take careful note of him, who he is and
where he lives, and bring the answer back to me.'

A long time passed before one of the Brahmans, Parnāda his
name, returned to Bhīma's daughter and gave his message as
follows:

'O Damayantī, as I sought Nala I came to Ayodhyā, and
though I there addressed Rituparna and his whole assemblage
many times, none answered anything. But when I was dis-

missed by the king and sat alone, one of his household, Vāhuka
by name, a deformed and short-armed charioteer, came and
sat by my side and spoke these words:

"That she was cast aside by an unfortunate,
A joyless fool, did not make her enraged with him,
His garment snatched by birds while hunting, burnt by fate;
Nor this could make the dark-eyed maid enraged with him,
When, treated well or ill, she sees her husband flown,
Of realm and fortune reft, and thrust from land and throne!"'

As soon as Damayantī heard this she hastened to her mother
and had Sudeva summoned. To Parnāda she gave rich reward
and promised richer still when Nala should return. Then to
Sudeva she gave the following command:

'Go to Ayodhyā and speak thus to Rituparna as if you had
come there by chance: "Damayantī, daughter of King Bhīma,
is holding her *svayamvara* once again; all kings and princes
hasten there; this time she chooses when the sun is rising with
tomorrow's dawn. If you would win her, hasten, for no one
knows whether Nala is alive or dead."'

Sudeva straightway went forth and sped over the distance
to Ayodhyā.

When Rituparna heard the words of Damayantī's messenger,
he called for his charioteer Vāhuka and thus addressed the
unknown Nala:

'I wish to go to Vidarbha and reach there by the morning,
for Damayantī holds her *svayamvara*.'

The unfortunate charioteer was distraught, uncertain
whether to believe completely in what he had heard or whether
to suspect some artifice on Damayantī's part. In the end, he
decided to carry out Rituparna's command in the hopes that
he might thereby also accomplish something for himself. Thereupon he entered the royal stables and, inspecting all the cour-
sers carefully, picked the four that seemed the best, four beautiful Sindh horses, a perfect blend of speed and strength.

As he drove them forward, Rituparna and Vārshneya sat and

wondered at his skill, Vārshneya debating within himself whether it was Mātali, charioteer of the gods, or whether Vāhuka was equal in skill to the once-famed Nala. These and many questions Vārshneya pondered as the chariot sped upon its way to Bhīma's city.

Rituparna then determined to show his charioteer his skill in numbers.

'Do you see,' he said, 'that *vibhītaka* tree? How many leaves would you say have fallen, one more than a hundred? And of fruits how many? And on those two branches, would you say, were there five times ten thousand leaves?'

Immediately Vāhuka checked the steeds and said: 'I cannot see what you say, O king, only let me stop and count while Vārshneya holds the steeds.'

But Rituparna answered: 'We have not time to delay.'

'Either stay a moment while I count,' requested Vāhuka, 'or else let Vārshneya drive you on to Vidarbha.'

But the king was loath to continue without Vāhuka at the reins, so he allowed him to count one special part of one branch as a trial. Whereupon he jumped quickly from the chariot and tore the branch away.

'You are right, O king,' he said, 'there are as many fruits as you said. I would give much to know the secret of this science; indeed if you will grant me this skill of yours, I will impart to you my skill in horses.'

Rituparna, anxious to reach Bhīma's city, agreed at once, and as his skill in dice passed into Nala, Kali flew out from Nala's body, vomiting Karkotaka's bitter poison. Nala would have cursed him, but Kali begged him, saying:

'Long have I dwelt in agony in your body through the serpent's poison. Now I pray mercy, O king, so that wherever men shall praise you in the world, no fear of me will remain in their hearts.'

With these words Kali, seen by none but Nala, entered the cloven *vibhītaka* tree.

Then as Nala finished counting the fruits, proudly he

mounted the chariot and, freed from Kali, still wanting though his own straight limbs, drove onwards to Vidarbha.

As evening was falling in Vidarbha, the sentries on the palace walls announced the arrival of Rituparna, and all within the palace heard the mighty rush of Nala's steeds. Vārshneya and Nala, still as Vāhuka in appearance, got down from the chariot and let loose the horses. Then Rituparna descended and went to meet Bhīma. Bhīma, who did not know of Damayantī's plot, bid the king welcome and asked him the reason for his visit. Rituparna, the embodiment of wisdom, when he saw no preparations for a festival, no signs of rival suitors, said no more than:

'To salute you I have come, O great Bhīma.'

But Bhīma too reflected that Rituparna had most certainly not set forth on a journey of over a hundred leagues merely to greet him. However, he too being wise, was content to wait for the matter to unfold of its own accord so, after entertaining his guest in regal fashion, he bid him good night and good rest.

No less puzzled than the two kings was Damayantī herself, for had she not heard the mighty trampling of Nala's horses, yet all she had seen was Rituparna with Vārshneya and the deformed Vāhuka. She called her handmaid Keshinī.

'Go, Keshinī!' she said, 'ask who is that short-armed charioteer. I doubt but whether he is Nala, yet be sure you speak just as Parnāda spoke before you come away and bring me back his answer.'

Keshinī hastened to fulfil her errand and approached the charioteer:

'Sent by Princess Damayantī, I salute you and would inquire from where you have set out and with what purpose you have come?'

'My master, king of Kausala,' Vāhuka replied, 'heard from a Brahman of your mistress's second *svayamvara* and so he has come, swifter than the wind, with me as charioteer.'

Keshinī then asked him who was his companion.

'That is Vārshneya, who was Nala's charioteer,' replied

Vāhuka, 'he retired to Ayodhyā when Nala fled away, whither, no one knows.'

Keshinī went on: 'You remember how the Brahman spoke that first came to Ayodhyā:

> "Where did you go then, you gambler,
>     And cut half my garment to keep?
> Why did you leave me, my loved one,
>     Your love in the forest asleep?
>
> "At your command she awaits you,
>     She stays there and waits for your words,
> Wasted away in her sorrow,
>     A girl whom a half-garment girds.
>
> "Silent she sighs in deep sadness,
>     Tear-stricken for you, her dear lord;
> Hero, please grant her a favour,
>     And to her some small answer afford!"

My princess would once again hear the words you uttered then.'

When the charioteer heard these words again his heart was shaken with grief and his eyes suffused with tears; with his voice choking, he repeated his reply:

'That she was cast aside by an unfortunate,
A joyless fool, did not make her enraged with him,
His garment snatched by birds while hunting, burnt by fate;
Nor this could make the dark-eyed maid enraged with him,
When, treated well or ill, she sees her husband flown,
Of realm and fortune reft, and thrust from land and throne.'

Then Keshinī took her leave, intent on relating to Damayantī the whole story and the manner of its telling. When Damayantī heard her maid's description, more than ever was she sure that it was Nala, but once again she sent Keshinī, bidding her:

'Keep careful watch over all his acts, but in particular see that he be given neither fire nor water with which to prepare his master's food.'

No great time passed before Keshinī returned with the wonderful tale:

'As he approached the doors, he did not stoop his head, indeed the portals flew up of themselves to grant him entry. When he came to prepare Rituparna's food, all the cooking vessels stood full of water as soon as he looked at them; nor did he need fire, for he took a mere handful of dry grass and as he held it towards the sun it straightway broke into flame. Not only did he handle fire without harming himself, but even the flowers which he picked up were refreshed after lying in his hands.'

When Damayantī heard this she was now convinced that her lord was there disguised in form as Vāhuka, but she sent Keshinī on a final errand.

'Go,' said she, 'and without being seen, bring some food to me that Vāhuka has prepared.'

Once again the worthy maid speedily carried out the task and brought, unnoticed, a morsel to her mistress. As soon as Damayantī tasted it, 'The charioteer is Nala!' were the words that tore from her lips, and forthwith she sent her two children with Keshinī to him.

When Nala saw them, he rose up swiftly and put his arms around them. Torn between the love of his children and the need to maintain his disguise, for he still thought that Damayantī was seeking another husband, the charioteer knew not what to do, when suddenly Damayantī appeared before him and asked him simply:

'What offence can I have given my husband that he should have abandoned me as he did, that he should have left me alone in the forest? Where were his marriage vows then?'

'Kali it was,' replied Nala, 'that brought me to this, not mine own deed; now that he has gone from me, here I am again. But how now does my noble lady choose a second lord?'

'Only a wile this,' Damayantī replied, 'once I knew from Parnāda where you were, for none but you could have driven

from Ayodhyā in a day. I call the spirit of the wind to testimony
that I have been true to you.'

As she spoke, the voice of the wind came from the space
around them:

'She is pure as when you left her, O king; we have watched
over her these three years. Know that this second *svayamvara*
was planned on your behalf alone. Put aside all treacherous
doubts and take your wife to your heart.'

Flowers fell about them through the words of the wind, and
the music of the gods was heard over their heads. Nala then,
bringing out the heavenly garment the serpent had given him,
put it on and instantly assumed his own noble form. Long the
two lay in each other's arms, long the two told each other their
stories.

# The Rāmāyana

*For as long as mountains shall endure and the streams flow,*
*so long shall the story of Rāma be current among men.*

## THE MARRIAGE OF RĀMA AND SĪTĀ

ON the banks of the Sarayū river, rich in money and grain, there lies the kingdom of Kosala, famed throughout all lands for its capital, Ayodhyā, built in ancient days by Manu, father of mankind. Once King Dasharatha ruled there, contenting his subjects and the law, and the kingdom prospered under his sway. Fulfilling all his duties, one thing only he lacked: a son to carry on his line.

Summoning his household priest Vasishtha, he requested him and the other great Brahmans to prepare a sacrifice. Whereupon all of them, with Vasishtha at their head, vowed to him:

'You shall most surely, your majesty, beget the sons which you desire.'

Turning to the ministers then, King Dasharatha said:

'Let the sacrificial ground be laid out forthwith; obey the priest's commands.'

When a suitable place was found, and the ground prepared, all according to the ancient precepts, Vasishtha and the other priests began the sacrificial acts. The sound of Vedic *mantras*, the sacred verses, filled the air, and there arose from the sacrificial offerings an incense which dispelled the sins of the king as he breathed it. When the fire god Agni appeared, there came with him the other gods surrounded by a host of *siddhas* and *gandharvas*, the heavenly saints and singers. With one voice they all pronounced:

'Since the demon Rāvana has appeared on earth and was given power by the Supreme Spirit, there is none of us free from his attacks, for we are unable to subdue him. May therefore the great Brahma, the creator of the worlds, grant that a mortal be born and that he may conquer Rāvana, the foe of gods and sages.'

Whereupon there arose, from the fire of the sacrifice, a spirit of matchless splendour, holding in his hands a vessel filled with heavenly fluid, and the spirit spoke to Dasharatha:

'Take this drink, most gracious majesty, and give it to your wives. When they have drunk, you will beget the sons on whose account you have made this sacrifice.'

The king then, overjoyed, accepted the vessel with bowed head and, returning to the palace, gave half the heavenly drink to Kausalyā and, of that left, half he gave to Sumitrā. To Kaikeyī then he gave half of the quarter over, and returned once more with what remained and offered it to Sumitrā. While Dasharatha was engaged in this task, Brahma commanded the gods who had returned to heaven:

'Beget sons who are your equal in valour but who have the form of monkeys!'

One year went by, and Dasharatha's wives gave birth to sons; Kausalyā first, her son was Rāma, while to Kaikeyī Bharata was born and to Sumitrā, two sons, Lakshmana and Shatrughna. Vasishtha, the household priest, was overjoyed and on the eleventh day after their birth he performed their name-giving ceremony. Of the four brothers, Rāma and

Lakshmana grew up in fondest friendship, while Shatrughna was ever attached to Bharata.

The time then came for King Dasharatha to decide whom should be given in marriage to Rāma and while he deliberated the matter with Vasishtha, there came to his court the mighty ascetic Vishvāmitra. Vishvāmitra, who by age-long penance had become a seer of the Brahman caste, although a warrior of the Kshatriya caste by birth, was welcomed by the king in manner fitting to his state and he then informed the king of the reason for his coming.

'In the course of my ascetic life, your majesty, my meditation and the religious rites which I perform are for ever being disturbed by two *rākshasa* demons, Mārīcha and Subāhu by name. Their valour is great, they are skilled in arms and they can assume any shape at will. My nature is such that I may not in anger contend with them. Therefore, your majesty, I demand the help of your eldest son Rāma, that he may destroy the *rākshasas* who assail me.'

When he heard this request, the king was for a while thunderstruck, but, regaining his composure, he addressed the noble ascetic:

'My son Rāma, with his lotus eyes and the down on his cheeks, is not yet sixteen years of age. My great army might perform the task that lies before you, but I do not see how Rāma can overcome these demons, for he has not even yet completed his training in warfare.'

Hereupon Vasishtha advised the king, with due regard for his feelings:

'Fear not, your majesty,' said he, 'that Rāma will come to any harm, for Vishvāmitra himself could overcome these *rākshasas*, skilled as he is in weapons. Let him watch over your son and complete his training.'

King Dasharatha was consoled at this and sent for his son Rāma as well as for Lakshmana, and with Kausalyā he performed rites to bring good and to ward off evil. Then, with Vasishtha's blessing, the three went off.

When they had gone a league or more along the banks of the Sarayū, Vishvāmitra called Rāma to him, saying:

'Come here, child! Take water in your hands and receive the spells for mighty strength. No one now will be a match for you in the three worlds.'

That night they passed on the banks of the Sarayū and the next morning, while Rāma and Lakshmana lay on their leaf couches, Vishvāmitra awakened them.

'Rise up!' said he, 'the new dawn draws near. Rise up and perform the daily rites of worship.'

So, when they had bathed in the river and made the purifying offerings of water, Vishvāmitra pointed out the forest that lay close by and told them:

'There, across our path, in that forest there lives the demoness Tātakā. She is the mother of Mārīcha, to slay whom I sought your aid. Now, Rāma, this demoness is a plague to Brahmans and cattle alike, so it is your duty to execute her, nor need you have pity because she is a woman. He who protects the interests of his people must always perform what is needful, whether harsh or kind, to uphold the law.'

Inspired by Vishvāmitra's brave words, Rāma, remembering how his father too had bade him obey such commands, took up his bow and, grasping it firmly, twanged the string with a fearful sound, so that the very air shuddered. When Tātakā heard that sound, she stormed in anger and, advancing through the forest, came face to face with Rāma. With a fearful cry she bore down on Rāma and Lakshmana, but Rāma pierced her with an arrow, and she died that instant.

Once Tātakā was slain, the two brothers went on with Vishvāmitra through the forest until, when the sixth day came, there appeared before them suddenly the two demons Mārīcha and Subāhu. In his anger Rāma hurled a radiant javelin at Mārīcha, and the blow struck so mightily that Mārīcha was lifted up by it and hurled far away into the ocean. Then, picking up a flaming club, Rāma flung it at Subāhu's chest, whereupon Subāhu straightway fell down dead.

Now that the *rākshasas* were slain, and the sacrifices no longer molested, Vishvāmitra conducted a sacrifice in Rāma's honour and declared:

'My purpose has been gained, and through you, mighty-armed Rāma. Now let us journey to Mithilā, the city of King Janaka. There you will behold King Janaka holding a sacrifice; there too you will see the mightiest of bows and one worthy of a hero such as you.'

From the battleground with the demons they set forth north-eastwards and ere long they reached Janaka's abode, the city of Mithilā in Videha. When Janaka heard that Vishvāmitra had arrived, he immediately came out, together with his household priest Shatānanda, to receive him. With hands folded and bowing deeply, he spoke:

'Most worthy ascetic, indeed I am honoured that you favour me with a visit and doubly fortunate in that you arrive when I am holding a sacrifice. But tell me, most noble sage, who are these two youths of heroic appearance who accompany you?'

Whereupon Vishvāmitra presented the two sons of Dasharatha to Janaka and recounted their adventures and told that they had come to find out about the bow. When Janaka heard this, he immediately replied:

'Here is that very bow of which you speak, and if Rāma is indeed able to bend this bow and string it, then I give him my daughter Sītā, born of the earth, and not of mortal mother.'

At these words of Janaka, one hundred and fifty men came forward, pulling behind them an eight-wheeled chariot, whereon lay the mighty bow. Rāma stepped forward and, picking up the bow, bent it so that it broke across and lay in two pieces, and at the noise the heavens shook.

'Now I have seen,' said Janaka, 'a strength and valour such as I have not believed to see. Rāma, son of Dasharatha, descendant of the solar race, is indeed a mighty hero. That my daughter Sītā should obtain him for a husband is an honour to my lineage. Now, most noble Vishvāmitra, with your consent my ministers should proceed forthwith to Ayodhyā and invite King Dasharatha here to Mithilā with all due courtesy.'

No great time went by before the emissaries returned and with them to Mithilā they brought the aged Dasharatha. Thereupon Janaka, approaching him, spoke words of greeting:

'Welcome to the land of Videha. Great honour is done to Mithilā by the presence of the great King Dasharatha. Now, with the highest contentment, I give you two daughters-in-law, Sītā as wife to Rāma, and Urmilā, a bride for Lakshmana; while for Bharata, and Shatrughna, let my brother Kushadhvaja present his daughters Māndavyā and Shrutakīrti, as brides to those inseparable brothers.'

King Dasharatha graciously accepted his daughters-in-law, and thereupon the marriage festivals were discussed and settled.

When the auspicious moment came, King Janaka brought Sītā, arrayed in all the ornaments of a wedding day, to the altar and, placing her face to face with Rāma in front of the sacred fire, he spoke these words to Rāma:

'This Sītā, my daughter, shall be your constant companion in all your duties. May you fare well, take her and hold her hand in yours, for she will be faithful to her husband, most blessed and ever following him as a shadow.'

Rāma then took her hand and pronounced the marriage vows and, with his brothers holding the hands of their respective brides, received King Janaka's blessing, and they all walked round the sacrificial fire, keeping it on their right hand.

The next day Vishvāmitra took his leave and went to the northern mountains. King Dasharatha too, when he had thanked Janaka for his most gracious hospitality and fine presents, set off on his return to Ayodhyā and with him went his four sons together with their wives.

Not far along the journey, the sky darkened and a fearful sound was heard. While Dasharatha was wondering what such terrible signs might mean, he suddenly beheld the son of Jamadagni, destroyer of the Kshatriya race, the warrior caste, approaching along his path. Of terrifying appearance, his locks matted, and bearing club and bow, he whose name was Parashurāma addressed himself to Rāma:

'I have heard of the great valour of Dasharatha's son. I have heard too how you broke the mighty bow. So now I have come here with the bow my forefathers obtained to overcome the warrior caste. Prove your strength therefore and fit an arrow to this fearsome bow that once was Jamadagni's.'

When Rāma heard these words of Parashurāma, he strung the bow and, fitting an arrow to the string, he drew it back. The son of Jamadagni stood there amazed as he watched the tremendous feat.

'Now I know,' said he, 'that you are indestructible, a slayer of demons and the equal of gods. Nor do I feel shame to have been humbled thus by one truly the lord of the three worlds. Release the arrow, for I yield up all my territories and I now make haste to Mahendra mountain.'

Rāma then let the arrow speed, and all at once the lands that owned Parashurāma's sway fell from his dominion. When King Dasharatha saw the successful outcome of the encounter, he rejoiced exceedingly, and thereupon, without any further mishap, they all continued the journey to Ayodhyā, where the four brothers lived with their wives, contented in serving their father.

The seasons passed for Rāma and Sītā, and she, for her gentle nature and great beauty, was beloved by him, the wife chosen for him by his father, while to her, Rāma was doubly beloved, for though his heart read what lay in hers quite openly, Sītā from Mithilā, daughter of Janaka, read his, as might a goddess, to the very deepest thought.

## THE BANISHMENT OF RĀMA

Once then, Bharata and Shatrughna, inseparable friends, went to stay with an uncle, Kaikeyī's brother, Ashvapati. While they sojourned there, King Dasharatha, who knew the time had come to pass his duties on, began to decide which of the four should succeed him to the throne, for all four were dear to him like four arms sprung from his body. Yet, of them all, Rāma, who brought happiness to all beings as well as to his father, was the best endowed with fine qualities. So Dasharatha assembled all his ministers and his court and addressed them:

'Now that I am grown old,' said he, 'the time has come that for the welfare of this land a crown prince should be chosen. He whom I consider foremost amongst my sons I now wish to appoint as *yuvarāja*, namely Rāma, for the upholding of righteousness in this realm.'

At this speech all the chieftains of the land raised up with one accord their lotus-like hands and exclaimed in unison:

'We desire Rāma the hero, Rāma of mighty arm and strength, to ride the royal elephant and shade his face under the royal umbrella. His care for the citizens is as that for his own relations. Let your son Rāma, then, Rāma dark as the blue lotus, Rāma conqueror of enemies, be installed as crown prince.'

Then Dasharatha spoke:

'In this auspicious month of Chaitra, when the groves are

flowering, let everything be made ready for the ceremony of installation.'

Then, turning to Sumantra, his charioteer, the aged monarch said:

'Go, bring Rāma quickly here!'

When Rāma appeared before his father, King Dasharatha told him of his appointment and spoke to him words of advice:

'All my people favour you, and you will therefore be made *yuvarāja*. Remember this always: that king who rules his country righteously and makes his people contented and loyal, all delight in him as the gods delight in nectar. So watch your conduct always and keep your senses in restraint. Tomorrow you shall be sprinkled with the water of consecration, so now return to your wife and pass this night controlling your thoughts and lying on a bed of *kusha* grass. Your friends should be ever watchful over you, for occasions such as these may be beset with difficulties. Bharata and Shatrughna are away from the city, but though, like their brothers, they are righteous and compassionate, I feel it may be better thus.'

When Rāma obtained his leave to go, he went to tell his mother Kausalyā and found with her Sītā as well as Sumitrā and her son Lakshmana. His mother wished him good fortune, whereupon he returned with Sītā to pass the night in his own house.

At that time a hump-backed slave of Kaikeyī, Manthará by name, mounted to the moonlit roof of the palace and saw the whole of Ayodhyā spread before her, festive with lights and dancing people and fluttering with flags and banners from every rooftop. She asked a nurse who was standing by what the cause of such merrymaking might be, and the nurse, in great delight, told her of Rāma's good fortune. At this the evil Manthará crept hurriedly back to Kaikeyī and there, giving way to her wrath, she exclaimed:

'Rise up, my mistress, this is no time to lie abed! Your husband may be righteous, but crafty and cunning is his nature.

First he sends Bharata away from the city and now, as I have learnt, he plans to instal Rāma as *yuvarāja* tomorrow.'

But Kaikeyī was not perturbed at this news and, handing her slave an ornament as a present, she said to her:

'What you have said to me, Mantharā, is pleasing to my ear, for I make no distinction between my son and Rāma. Rāma is the elder and therefore entitled to succeed his father, and indeed I cherish him as much as I do Bharata.'

But the wicked Mantharā refused to accept the gift and went on to vent her evil thoughts.

'If Rāma becomes king,' she cried, 'Bharata will have no place in the royal line. Do you not also see that when Rāma becomes king, his mother will wreak vengeance on you, for you surpassed her formerly by your comeliness in King Dasharatha's favour?'

'What do you propose then that I should do?' Kaikeyī asked.

'Go before the presence of the king,' said Mantharā, 'and, clad in dirty rags and making yourself as if enraged, remind the king of how you rescued him when he was wounded in battle. Remind him how he then promised to grant you two favours whenever you should ask him. Now the time has come to ask and this is what you should say: "Let Rāma be banished to the forest for nine years and five. Let Bharata, your majesty, be made ruler of the land." Thus, once Rāma has been banished for fourteen years, your son Bharata will have secured his position in the kingdom.'

To this most foul device Kaikeyī lent a willing ear and, giving her slave an abundance of ornaments, she lay on the floor in her apartments, feigning anger and distress.

When King Dasharatha had made all the arrangements for the installation and had ordered the royal sprinkling with water to be carried out for Rāma, he entered Kaikeyī's beautiful apartments, like the moon entering a cloud-fleeced sky where Rāhu, demon of the eclipse, lies in wait. When he saw his fair young wife lying on the ground, angry and unkempt, he caressed her, saying:

'What means this sad state? Tell me what has grieved you. I swear by him whose presence is ever a joy to me, I swear by Rāma that I shall do whatsoever you wish.'

Kaikeyī, then, reminding him of the favours he once had promised, spoke to him words that were cruel as the onset of death:

'At the royal ceremony let my Bharata be sprinkled with the water of royalty, and for nine years and five let Rāma wander as an ascetic.'

The king broke into a fire of rage and burnt her, as it were, with his eyes.

'You wicked woman,' he exclaimed, 'what sin has Rāma, what sin have I committed against you, cruel wretch? Rāma treats you as his mother, why should you seek to destroy him? When the praise of his virtues lies on every lip, what fault of his can I point to and say: "It was for this that I forsook him"?'

This and more the king spoke, half pleading as the torrent of his words ran out, but the terrible Kaikeyī remained adamant in her resolve.

'You have made your promise,' she said finally, 'and you must abide by *dharma*, the law of righteousness. For love of *dharma* and at my behest, you must banish your son Rāma. Three times I say this to you.'

King Dasharatha, knowing then that he could not make her abandon her vile request, rose up and proudly bade her farewell.

When the next day came, the household priest, Vasishtha, went into the city bearing the vessels for the royal ceremony. Sumantra, the charioteer, came up to him, and Vasishtha bade him go and convey King Dasharatha to the ceremonial ground. Reaching the king's bedroom, Sumantra announced himself and spoke to the king:

'As the sound of the Vedas, the sacred hymns, wake Brahma the self-born lord at this very hour, so now, your majesty, do I awake you.'

The king's only reply was:

'Charioteer, bring Rāma to me!'

Sumantra swiftly did as he was bid. Rāma then, arrayed for the ceremonial occasion, took leave of Sītā and, mounting the chariot amid the joyful acclaim of his friends, set out to his father's palace. Entering the audience chamber, he found King Dasharatha, his face sad and careworn, seated beside Kaikeyī. First he bowed before his father's feet and then he greeted Kaikeyī as well.

'Surely I have done some wrong,' said he then, 'for my father appears angry.'

Then, turning to Kaikeyī, he continued:

'Please tell me who ask you, what causes this most unexpected change in my father's countenance?'

King Dasharatha remained silent in his grief, but the cruel Kaikeyī wanted no time to declare what lay in her heart.

'Rāma dear, the king is not angry. It is just that he does not care to say to you what is unpleasant to hear. Yet, what he has promised me, that you must obey.'

With this, Kaikeyī went on and told him what had passed between herself and his father.

'So now you must forgo the royal ceremony,' she concluded, 'and live in the Dandakā forest for twice seven years, wearing the matted locks and the black deerskin of an ascetic.'

When Rāma the conqueror of enemies had heard these words, he overcame their death-like hurt.

'Let it be thus,' he said to Kaikeyī, 'I will go from this place into the forest, wearing matted hair and the deerskin, and so the king's promise shall be kept. I am only grieved that the king did not himself tell me of Bharata's succession, for I would willingly give up the kingdom and my life for Bharata.'

Thereupon Rāma left his father and Kaikeyī and went to see Kausalyā, honouring, with kind words, those who spoke to him on the way, for the loss of a kingdom does not affect one of true heart. He embraced his mother and told her somewhat hesitantly of what had come to pass. Great was her sorrow that the day she proudly had awaited was filled with such bitter news.

Lakshmana too approached and when he heard the cause of Kausalyā's lament, he was moved to violent words, but Rāma gently admonished him, saying:

'Son of Sumitrā, that I am banished and my kingdom taken is but the working of destiny, for how could Kaikeyī wish me harm or hurt? Her resolve must have been inspired through some divine decree. Who can fight destiny whose plan is seen only in his deed? Do not give way to grief or anger, my brother Lakshmana, when fortune goes against us. Between ruling a kingdom or living in the forests, the life of an ascetic has much to commend it.'

'Why do you hold destiny responsible for this?' asked Lakshmana. 'Can you have no doubts about the foul deeds of Dasharatha and Kaikeyī?'

But they saw that Rāma was bent on obeying his father's command, and so his mother Kausalyā said to him:

'Rāma, my son, how can I continue to live with Kaikeyī and Sumitrā? Take me with you to the forest if indeed you are determined to carry out your father's wish.'

'Bharata will care for you,' answered Rāma, 'for he is kindly to all; and remember this, so long as a woman lives, her husband is lord and master, so you must stay by my father's side. Even if a woman ceases to worship the gods, she gains the supreme heaven by obeying her husband.'

So, holding back her sorrow and touching holy water, Kausalyā gave her son auspicious blessings. Thereupon Rāma took his leave of her and returned to his own dwelling, where the flags still fluttered and the people thronged in festive joy.

Sītā came forward to meet him, nor did Rāma need to tell her that he brought with him harsh tidings. A comforting embrace, and then, all that had lain pent up within his heart broke out.

'Sītā,' he said, 'my father whom I revere has exiled me to the forest.'

Then, in his sorrow, he told her of all that had befallen and that she must stay honouring his father, devoted to his mother

and practising vows and fasts. Sītā listened quietly, but not without annoyance because of her great love, and then she said to
him:

'Why do you speak what you know to be impossible? A wife
has many duties, but the highest is to her lord. That you have
been commanded to dwell fourteen years in the forest is a command to me also to do likewise. If you are to leave for the ascetic
life this day, I shall walk before you, making smooth the sharp
grass and the thorns.'

Though Rāma tried to dissuade her, and told her how life in
the forest was beset with dangers and how those who dwelt
there often knew the pangs of misery and hunger, Sītā remained
settled in her resolve, yet grief-stricken lest he should leave her.

'Rāma, you know this true teaching of our holy Brahmans,'
said she, 'that a woman who, according to the law, has been
given by her parents to a noble suitor, remains his both in this
world and in the world to come. Rāma, I am devoted to you
and I must accompany you.'

Thereupon Rāma embraced her in his arms and soothing her
sorrow promised to take her with him.

When Lakshmana heard then what they had decided, he
said:

'If you have both resolved to enter the forest, full of wild elephants and other beasts, then I too, shall accompany you
armed. Bharata will be here to watch over our parents, so I
may become your follower.'

Rāma relented, and when Sītā, now full of joy, had finished
making her farewell arrangements and giving gifts, the two
brothers went along with her to see their father. Ushered by
Sumantra into his presence, they beheld him with his three
wives in attendance. He rose up and came forward to meet
them, but ere he had crossed the hall the aged king sank to the
floor under the burden of his great distress. Rāma and his
brother ran forward to help him to his feet.

'Now I must take my leave,' said Rāma to his father, who had
recovered somewhat from his faint, 'and though I have tried to

dissuade them, Sītā and Lakshmana want to go to the forest with me. Please then permit them to take their leave also.'

Whereupon, after many sorrowing words, Dasharatha bid them farewell, giving Sītā gifts of clothes and ornaments for her long sojourn in the Dandakā forest. The brothers made their farewells to their mothers, and then, going outside with Sītā, they mounted Sumantra's chariot. Before Sumantra whipped up the chosen steeds, Rāma spoke to the assembled citizens:

'The regard and love which you of Ayodhyā have shown to me, please now show them, even more abundantly, to Bharata. Do this and it will please me.'

Then, with swift pace, Sumantra brought them to the banks of the Tamasā river, and, when they had performed the twilight prayers, they went onwards to reach the boundary of the kingdom. Soon they came to Shringaverapura on the Ganges and there they met King Guha with his family; there they remained that night. The next day Rāma said to Sumantra:

'Return now to Ayodhyā, that Kaikeyī may know that I have gone to the forest and that she may not doubt my father's word.'

Then, bidding Sumantra and King Guha farewell, Rāma, together with Sītā and Lakshmana, crossed the River Ganges and, after journeying pleasantly awhile, they came to the meeting of the Ganges and the Yamunā, where, at evening-time, they met the ascetic Bharadvāja.

'I have heard,' said he, 'of your sad exile and have been awaiting you. Now I would tell you to travel to Chitrakūta mountain, a fair place, with sweet honey, roots and fruits, whereon you may live.'

Taking the ascetic's advice, they journeyed to Chitrakūta, and there Lakshmana soon built a pleasant arbour where the three of them dwelt, observing the vows of ascetics.

In the meanwhile Sumantra had returned to Ayodhyā and there, entering the king's presence, he gave the king a humble message from Rāma. Dasharatha could not restrain his grief, and lamenting said:

'Alas, that I should have let that Kaikeyī, her of sinful race, drive me to this most sinful deed. Wretch that I am, that I did not seek the counsel of my ministers, foolish that I let myself be moved to this through love of a woman.'

With these and like laments, the aged king's heart cracked and his life came to an end. When the ministers saw that the king was dead, they sat in solemn consultation with Vasishtha at their head. On his advice they resolved that Bharata should be brought back swiftly from Rājagriha, that the funeral rites might be observed and that Ayodhyā should not remain without a king. Forthwith the messengers set out and reaching Rājagriha, they gave this message to Bharata:

'Vasishtha and the ministers inquire for you. Please come back with us speedily for a difficult task lies before you.'

At this, Bharata, accompanied by Shatrughna, took his leave of his relations and though he knew not the reason for his journey, disturbing thoughts assailed him. On the seventh day he came to Ayodhyā but, not finding his father in the palace, he sought out his mother, whereupon the wretched woman told all that she had contrived. Foolish in her delight, she little imagined how her son would answer her.

'Treacherous woman,' said he, 'what do I want with a kingdom; bereft at one foul stroke of a father I loved and of a brother who was as a father? You have come to destroy our family. How could you not know that kingly Rāma is the refuge for us all?'

As the waves of his grief and anger rolled on, Vasishtha, ever wise, stemmed the restless tide and said in gentle voice:

'Enough of grieving! Now is the time for you to think of your duties to your father and to perform his funeral rites.'

At these solemn words, Bharata grew calm and he performed all the rites due to the dead, and on the twelfth day he made the sacrificial offering to the spirits of their forefathers. This duty done, the ministers prepared to instal Bharata as king, but Bharata refused.

'Let a force be prepared; bring my chariot too, for I intend to seek out Rāma, wherever he may be, and bring him back to his rightful inheritance.'

Whereupon they all set forth, ministers, priests and even the three mothers, all roused with excitement at the hopes of Rāma's return. When they reached the borders of the Dandakā forest, Bharata ordered his troops to enter the unwelcome shades, and thereupon such a clamour of frightened beasts and wild elephants set up that the sound of it reached the dwelling-place of Rāma. The fiery Lakshmana swiftly climbed a tree and from the topmost branch espied the spreading net of Bharata's men.

'It is Bharata come to hound us down,' cried he. 'Now that he has gained the kingdom he wants to ensure his place by putting an end to us.'

But the sage Rāma quietly calmed him from such violent thoughts.

'Bharata loves his brothers,' said he, 'and I am sure that he has come to see the three of us out of affection and for no other aim, except that he is displeased with Kaikeyī's conduct and wishes to offer me the kingdom.'

It was not long before Bharata reached the leafy arbour and there, in sorrow so great he knew not how to tell it, made known to Rāma that their father was dead. When Rāma heard the news, he fainted at the bitter shock, and when he recovered, went slowly with the others to the Ganges and there he too performed his part of the rites to the soul of the departed king. Thereupon they all returned to the hermitage, and there Bharata urged his brother in earnest words:

'Let me take your place here and wear the ascetic's deerskin. Your widowed mothers need your protection. True, the kingdom was given to me, but now I give it back to you.'

'Not thus, Bharata,' replied Rāma. 'Your mother asked two blessings of our mighty father, the kingdom for you and exile for me, and he, being bound, granted her these blessings. Therefore I am here at our father's command; do you also obey that command. Return to Ayodhyā and rule our subjects contentedly and righteously.'

Vasishtha, too, attempted to persuade him:

'Your father's teacher was I, as I was yours,' he spoke; 'by following the course whereby the eldest son in the royal line becomes king, you will do no wrong. A man has three preceptors, his father, his mother and his teacher; his father gives him birth and his teacher gives him wisdom. I tell you this and you do no wrong.'

'What parents do for a child,' answered Rāma, 'in giving all they can, by bathing him and watching over him with fond words, all that is not easy to requite. What my father has wished then, that I will do. You may see the moon bereft of beauty or the snow flown from Himālaya or the waves of the ocean leave the shore, but never will I abandon the promise to my father.'

Though Bharata entreated him the more, Rāma bade him not to brood on the outcome. Then Bharata, knowing that his brother's will could not be broken, put a pair of golden sandals before Rāma and asked him to step into them. Rāma did as his brother asked and, when he had stepped out of them, Bharata bowed low before them, saying:

'For as long as your return is awaited, I shall remain leading the life of an ascetic, living in Nandigrāma, outside Ayodhyā, and as for royalty, let these sandals be the symbols of authority. But if you do not return, Rāma my brother, in the fourteenth year, then assuredly I will enter the funeral pyre.'

Rāma promised that he would return and, bidding him care for their three mothers, said farewell to him and Shatrughna. Bharata then placed the sandals upon his head and returned to Ayodhyā. There he left his mothers and, installing the sandals in the palace after sprinkling them with the water of royalty, he went to Nandigrāma, from where he administered the kingdom.

But for Rāma, the hermitage of Chitrakūta was now full with memories of his sad relatives and subjects. With Sītā and Lakshmana he therefore journeyed to the hermitage of the great sage Atri and was received there as a son, and Atri's wife made

Sītā a most charming welcome, giving her many gifts. The next day they entered deep into the forest of Dandakā, after Atri had pointed them the way.

### THE LOSS OF SĪTĀ

When Rāma, with Sītā and Lakshmana, entered the Dandakā forest they came soon to a hermitage where the ascetics received them with every kindness and begged Rāma to protect them from the demons that infested the forest. Nor was it long before they were to meet one, for as they set out the next morning, there bore down upon them a man-eating giant. Swiftly he tore Sītā from her husband's arms and threatened the two brothers with death.

'First tell me who you are,' said he, and when the brothers informed him of their names, he continued: 'Know me to be Virādha and that I enjoy the Brahma-given power that no mortal may slay me with weapons.'

Then a fight began that lasted long in vain, for the giant's strength was such that he placed the brothers on his shoulders and bore them away deep into the forest, while Sītā followed them weeping. At last, by dint of mighty effort, Rāma broke the giant's one arm and Lakshmana the other. As Virādha sank down, Rāma put his foot on his neck and made him unconscious. Thereupon they dug a pit to cast the giant's body in and as they did so, a voice came forth from the body:

'In my true form I am the *gandharva* Tumburu, cursed to remain in this shape until slain by Rāma. Now you must proceed one and a half leagues from here to the hermitage of the sage Sharabhanga.'

With these words the spirit of Tumburu fled to the upper air, and the brothers cast the corpse into the pit.

When they reached the hermitage of Sharabhanga, he received them most courteously, surrounded by numerous ascetics. As they approached, Indra, chief of the gods, who was visiting Sharabhanga, disappeared in his chariot into the skies.

'I was about to depart,' said the venerable sage to Rāma, 'and go to the world of Brahma, but first I wished to await your arrival and to greet you. Now you shall continue on your way, following the holy river, until you find the sage Sutīkshna.'

Sharabhanga then instituted a sacrifice, and as the ceremony proceeded, he mounted the sacrificial pyre, whereupon there arose from the flames the form of a handsome youth which sped upwards to Brahma's heaven. Again the ascetics implored Rāma's aid against the attacks of the *rākshasas*, and Rāma promised them he would fight such demons with force of arms wherever he encountered them. Taking their leave then, the three wanderers followed the river bank and moved onwards towards Sutīkshna's hermitage.

Sutīkshna was expecting them, for Indra had warned him of Rāma's arrival.

'Welcome,' said Sutīkshna to the weary travellers. 'Rest you here awhile; gather up your strength for your battles against the demons, but beware for ever of the gazelles that sport among the forest trees, for they will bring you nothing but misfortune.'

One more night then Rāma passed with Sītā and Lakshmana, before they set out again the following day at daybreak. As they went along, Sītā turned to her husband, saying:

'It is true that these ascetics are assailed by demons, and indeed have shown us the bodies of their slain. True too that you and Lakshmana have pledged the strength of your arms in this cause. Yet it is said, nor here do I teach you, but would remind you only, that to kill without provocation is one of the three great sins born of passion, of which the other two are the uttering of falsehood and coveting another man's wife. These two are not your nature, how now will you embark upon a third? You are under the vows of an ascetic, and what place has the deadly bow of the Kshatriya caste among the frugal possessions of an ascetic?'

At these words, Rāma thoughtfully replied:

'What you say is true, dear Sītā, but I have given my promise to these ascetics who implored me, and rather would I abandon my life than swerve from the truth of a promise made, especially to Brahmans.'

Then, as they journeyed on, they came to a lake around which the air was filled with sweet music. While they were wondering at the heavenly sounds, a seer approached them and told them how a certain ascetic, Māndakarni, had been led to temptation there by five *apsarases*, heavenly maidens, wherefore the lake was known as Panchāpsaras, and that Māndakarni now lived in a palace on an island in the centre of the lake. The singing of those *apsarases*, went on the seer, caused those sounds which fell so sweetly upon the ear. There then, in the hermitages on the shores of Panchāpsaras, Rāma, together with his wife and brother, passed the next ten years.

When the ten years had passed, Rāma, ever mindful of his promise to aid the ascetics against the attacks of the *rākshasas*, went once again to visit Sutīkshna. Sutīkshna told him to seek out that redoubtable sage Agastya, since he had done battle with the demon brothers Ilvala and Vātāpi; for, so the story went, when Brahmans held their sacrifices, Ilvala would place the flesh of his brother, Vātāpi, among the offerings of food. When the Brahmans had eaten the offerings, Ilvala would summon his brother to life again, whereat Vātāpi would rend the Brahmans asunder as he stepped forth. But when Agastya ate of the flesh one time, Ilvala, try as he might, could not revive his brother, whereupon Vātāpi perished utterly, and Ilvala himself was slain by Agastya in the fight that followed.

As they entered Agastya's hermitage, the heroic sage greeted his fellow-hero Rāma most cordially and, inviting him to be seated, spoke these words to him:

'Your deeds make you a most welcome guest. Receive from me this mighty bow, of heavenly craftmanship and studded with jewels, and with it victory shall be yours even as Indra slew the demon Vritra with his thunderbolt. Two leagues from here there is a fair place, endowed with all the bounty of nature and known as Panchavatī, near the River Godāvarī. There make your abode with your brother and carry out your father's commands.'

As they travelled onwards then, they encountered a mighty vulture, huge in body and of fearful strength. The bird told Rāma that he knew of him well and that Dasharatha had been his friend.

'Know then, Rāma,' said the great bird, 'that I am Jatāyu.'

When they had exchanged greetings, they continued their way to Panchavatī, and there, by the banks of the Godāvarī, Lakshmana swiftly made a pleasant dwelling for his brother and Sītā.

One day, while Rāma was living happily with Sītā in that delightful place, there came a demoness, the *rākshasī* Shūrpanakhā, wandering by. When the ugly *rākshasī*, with nails as large as winnowing baskets, caught sight of Rāma's lotus face, she was immediately smitten with love for him and said to him:

'Why have you and your wife come to these parts, where only the *rākshasas* live, and how is it that though you are dressed as an ascetic, you carry bow and arrows? My name is Shūrpanakhā; I am a *rākshasī* who can assume any form at will. How is this woman suitable for you; if you will become my husband, I shall eat her up, and then we may wander happily together without disturbance.'

Rāma made no more reply than to send the unpleasant wretch to his brother. Lakshmana, of quicker temper, swiftly

rewarded the ugly *rākshasī's* demands by cutting off her ears
and nose, thus saving Sītā from a horrid fate, for Shūrpanakhā
was preparing to attack her.

Enraged and weeping, Shūrpanakhā hurried back in amongst
the forest trees and sought out her brother, the demon Khara.
When Khara learnt the cause of her distress he sent forthwith a
band of fourteen *rākshasas*, with Shūrpanakhā, thirsting to
drink the blood of Sītā and Lakshmana, to show them the way.
Yet when the demons arrived at Rāma's hermitage, it seemed
a miserable force to send against such a mighty hero. Before
they ever reached the grove where Rāma's abode lay, the fourteen lay tangled in the forest grass with Rāma's arrows through
their hearts. Once more Shūrpanakhā sped back to her brother
and, telling him of the *rākshasas'* defeat, entreated him to make
one more great effort to avenge her.

'Come, Dūshana,' said Khara to his fellow-*rākshasa*, 'let us
summon all our forces, fourteen thousand men.' So Khara
mounted his chariot and drew out from Janasthāna with the
tumultuous army of demons. Evil omens filled the sky, but
Khara's demons marched on to Rāma's hermitage nonetheless.
At the prospect of the fearful battle that was to come, the gods
and the *rishis*, the divine sages, looked down out of heaven and
thought upon the outcome. As Khara approached the hermitage, he saw Rāma, angry in countenance and making ready his
mighty bow. Yet, though the demons in their wrath rained
arrows upon Rāma, their aim was wild in their excitement, and
Rāma, unscathed, slew so many that the rest withdrew. Victorious, he stood alone, for he had placed his brother and Sītā
in the shelter of a cave. Then Dūshana reassembled the army
of *rākshasas* and once more they rushed forward like stampeding
beasts out from the trees of the forest. Rāma plied the *gandharva* weapon against them and thousands of *rākshasas* fell
dead. Dūshana himself charged up to Rāma in his chariot but
one of Rāma's arrows pierced the chariot through and killed the
demon in it. Soon of all the *rākshasas* there were but Khara and
Trishiras left. Trishiras rushed forward to the attack and met

Dūshana's fate. Thereupon the mighty Khara let fly his arrows
at Rāma so that Rāma's bow was sprung from his hand and his
shield flew in pieces. Swiftly Rāma ran to where the bow which
Agastya had given him lay, and as he fired his arrows swifter
than hailstones, Khara's chariot fell asunder and his left hand,
together with his bow, was shot away. In terrible rage, Khara
flung his club at Rāma, but long before it reached him a well-
aimed arrow stopped its flight in mid-air. Then Khara tore out
a great *shāla* tree by the roots and hurled it. This weapon too,
Rāma foiled with a hail of arrows and soon, as the *rākshasa's*
strength was exhausted, Rāma finally laid him low. So the
hideous realm of Janasthāna was utterly destroyed and the
*rishis* appeared in heaven, praising Rāma for carrying out the
glorious deed.

In the meantime Shūrpanakhā, who had witnessed Rāma's
tremendous feat, hastened to the island of Lankā, where her
brother Rāvana reigned supreme. There in Lankā, on the high
terrace of his palace, Rāvana sat in council with his ministers
when Shūrpanakhā burst in amongst them. She told Rāvana
swiftly how the dominion of the demons had been destroyed by
Rāma, and Janasthāna cleared of their kinsmen. Lamenting she
told how she herself had been disfigured by Rāma's brother.
Finally she told Rāvana of Sītā's rare and wonderful beauty and
how whoever should possess her would gain kingship over the
whole earth.

Straightway Rāvana had his chariot prepared and hastened
to the shore. Then, crossing over the ocean, he met Mārīcha,
he whom Rāma had hurled into the sea, dwelling on the other
shore. When Rāvana told Mārīcha of the fearful fate their fellow demons had suffered at Rāma's hands and how he intended
to slay Rāma and carry Sītā away, Mārīcha spoke to him:

'You could not understand Rāma, for he is noble in character, and you are base, nor can your schemes succeed. What use
were your ministers and your spies that you did not know what
was happening in Janasthāna? And further, what more foul act
than to take another's wife? You say that Rāma disfigured

Shūrpanakhā and killed your brother and Dūshana and fourteen thousand *rākshasas*, all without provocation, but I think
there was some fault with them, and I know Rāma's might.
Fight not with him.'

'No, Mārīcha,' Rāvana replied. 'I am settled in my resolve,
nor could all the gods make me swerve from this. You do not
know your place that you should speak to me thus. Now I command your help and this is what I will have you do. You who
can assume any form at will, shall transform yourself into a
golden deer and roam about in Rāma's hermitage grove where
Sītā may see you. Then Sītā will ask for you to be brought to
her, and when you hear this, you shall flee far away and then
cry out: "Ah, alas Lakshmana, alas Sītā!" Do this now nor
disobey, else I shall kill you.'

Whereupon they sped off in Rāvana's chariot towards Panchavatī, and Mārīcha did Rāvana's bidding. When Sītā saw the
beautiful golden deer cropping gently in the leafy shade of the
hermitage grove, she called to her husband and to Lakshmana.
But they on seeing the creature were forced to doubt. Lakshmana turned to his brother and said:

'It would not surprise me were this that very demon whom
once you hurled away but did not kill, Mārīcha, come to play
us an ill game.'

But Sītā was so captivated by the loveliness of the animal that
at last she won her way, and Rāma fell in with her wishes.

'Do you remain here!' he ordered Lakshmana, 'nor move one
instant from Sītā's side while I go in quest of this deer.'

Then, drawing a circle in the earth around Sītā, Rāma put on
his sword and, taking his bow, set off in pursuit of the deer.
Mārīcha, still in the form of the deer, drew him even farther from
the hermitage. Finally Rāma determined to kill the animal,
and as Mārīcha, transfixed by a deadly shaft, sank upon the
ground, he called out: 'Ah, alas Sītā, alas Lakshmana!'

When Sītā heard that sad cry, which seemed indeed her husband's, she forthwith begged Lakshmana to seek what had happened. But Lakshmana, mindful of his brother's command,

refused to go. At last, in tears and rage, imploring him again and again, Sītā persuaded him to leave her side and find out whether her husband had come to harm.

No sooner had he gone than a wandering ascetic appeared before Sītā. In sly words the beggar asked her:

'Who are you, that one so beautiful should be found here in the forest, dressed in gold and silken cloth and wandering at will?'

Sītā told him who she was and how she had come there to live with her husband and his brother. Then, as she stepped outside the circle drawn on the ground and went to bring the wandering beggar some alms, immediately he said:

'Know me to be the famous ten-necked Rāvana, lord of the island of Lankā. Become the wife of one who is known to all the three worlds, become my bride and leave this mere mortal!'

With these words Rāvana assumed his true form and, seizing her by the hair, he caught her up in his arms and led her away in his chariot. Lamenting in her distress did not avail Sītā, borne on in the rough grasp of the king of the *rākshasas*, but then she saw Jatāyu and called out to him to tell Rāma of her fate. The noble vulture swooped down and, barring Rāvana's path, challenged him to battle. But alas for Jatāyu that his friendly service and great courage helped him not, for Rāvana with all his ferocity quickly drew his sword and cut off the vulture's wings and legs. The unfortunate Jatāyu lay helpless and bleeding on the ground while the dust of Rāvana's chariot rolled far away. As the chariot sped, Sītā cast her flowers and ornaments on the ground and once, as they passed through the sky over a mountain-top, she beheld five powerful monkeys squatting on the peak. To them she cast her golden scarf and some ornaments besides, hoping perchance that they might tell her husband.

Soon they approached the island of Lankā, and Rāvana brought Sītā to his palace, yet, despite his attempts to win her and make her his bride, Sītā remained with her heart as stone to his entreaties, her face veiled in shame. At last the angry demon swore that if she did not change her mind within twelve months, he would eat her alive. With that he put her to live in an *ashoka* grove, guarded by *rākshasis*, but the gods in heaven rejoiced, for they knew that Rāvana's defeat would come.

When Rāma had killed Mārīcha in his form as a golden deer, he returned hurriedly to the hermitage, but as he ran, he met Lakshmana on the way. Furious with his brother for not remaining with Sītā he flew back to where he had left her. Empty was the arbour, empty too all the caves and clefts and forest strongholds that Rāma, heartbroken in his love, searched with Lakshmana. Then, as they spread out ever farther from the hermitage, they suddenly found poor Jatāyu, near to the point of death.

'Alas, dear Rāma,' said the sorely stricken bird, 'Sītā whom you seek and my life as well, both have been carried away by the foul demon Rāvana. I fought as I could, but when he left me in this helpless state, his chariot rolled onwards to the south. Alas, Rāma, my life will soon end, and before my eyes I see trees with golden leaves and branches. Despair not, noble Rāma, for in that moment when Sītā was taken from you, in that moment is the seed of your finding her again.'

With these words, Jatāyu, most noble lord of birds, breathed out his life. Rāma lamented bitterly the loss of such a true friend, and then, making a funeral pyre, he burnt Jatāyu's body to the accompaniment of all due rites, as he would his own kinsman.

The brothers wandered onwards then to the south through the great Krauncha forest, wild with huge trees and fierce animals. Then suddenly there arose the sound of fearful crashing as of some mighty beast breaking through the undergrowth. In front of them appeared the mighty Kabandha, a headless monster with vast outstretched arms. So huge was he that Lakshmana quailed and Rāma too thought that his final moment had appeared. Then, as the monster approached to destroy them, they found courage in their terror and wielding their swords, hewed off its arms. Brought to helplessness the monster asked them who they were and when they told him he then began to relate his own story.

'Kabandha am I, who took this form to frighten all the *rishis*,

the great sages,' said he, 'but the *rishi* Sthūlashiras cursed me
so that I would remain in this shape until Rāma and Lakshmana cut off my arms. I am the son of Dānu, gifted with age-
long life, which made me struggle against the great god Indra
himself, but Indra cut off my head and swore that I should
remain like this till Rāma should redeem me. Now, I pray you,
burn me in a pit and I will tell you news of how you shall reach
Sītā.'

As the flames then licked about the monster's trunk, there
arose a radiant figure from the fire which spoke to them:

'Go now swiftly and seek out the valorous Sugrīva. He
dwells by the Pampā lake. Go make alliance with him, for
Vālin, Indra's son, has robbed him of his rulership over the
monkey tribe. Gain his help and he and his hosts will soon find
Sītā for you.'

Thus heartened, the brothers sped quickly along the path
Kabandha had pointed out and not long time passed before they
reached the Pampā, the very beauty of which, with water-lilies
and lotuses of many colours, made Rāma's grief burst out anew.

THE MONKEYS' SEARCH FOR SĪTĀ

As Rāma lamented and wondered how he would tell King
Janaka of Sītā's loss and how all his family would grieve to learn
that he had lost his wife, Lakshmana comforted him,

'Now is not the time for sentimental grief,' said he, 'but rather for action.'

Heartened thus by his brother's words, Rāma proceeded across the Pampā lake, swathing through the lotus clusters that bowed their stems in welcome. As they approached the other shore, Sugrīva espied them and, fearful at their valiant stature and formidable weapons, he withdrew speedily to Malaya mountain, saying to his lieutenant Hanuman:

'Noble monkey, find out whether these two have come with peaceful aim.'

Hanuman forthwith approached the brothers and, bowing courteously to them, addressed them with these words:

'Why have you, of most comely appearance, come to these parts? And may I know too who you are? My master is Sugrīva, a brave and righteous lord who wanders about in sorrow, bereft of wife and kingdom by his brother. Hanuman, a monkey, his minister am I, sent by him to inquire after you, for the noble Sugrīva desires friendship with you. Know that I am the son of Vāyu, god of the wind.'

When Rāma heard these words he turned to Lakshmana and said:

'This is the minister of that very king of the monkeys that Kabandha told us of. Speak courteously to him, Lakshmana, for indeed Hanuman would not speak as he does were he not versed in grammar and the hymns of the Veda.'

'Indeed we seek this Sugrīva,' Lakshmana then addressed the noble monkey, and after he had told him of all the adventures which had brought his brother and himself to Pampā lake, he continued, 'We know not where the demon is who has carried Sītā away, nor where he dwells, so take us therefore to Sugrīva who is our refuge.'

At this the mighty Hanuman, son of Vāyu, took the two heroes upon his back and carried them straightway to Malaya mountain. There he announced them to Sugrīva and informed him who they were. Sugrīva was greatly pleased at their coming and welcomed them:

'Since you who are noble and righteous seek friendship with me who am a monkey, all honour and the highest gain accrue to me. If you indeed wish this alliance, then here is my hand, please take it in token of our understanding.'

Rāma gladly accepted the proffered hand, and together they walked round the fire, keeping it to their right. Then Sugrīva said:

'My minister Hanuman has told me of your story and how your wife was carried away. You may be sure that I shall bring her back, for I saw her when the demon bore her aloft in his chariot, crying: "Ah, alas Rāma, alas Lakshmana!" She dropped a golden scarf and some ornaments; here they are.'

When Rāma saw those possessions of his beloved wife, he lamented bitterly. 'Look, Lakshmana,' said he, 'see these things of hers.'

'Indeed,' said Lakshmana, 'I do recognize those anklets, for often have I bowed before her feet.'

'Tell me, Sugrīva,' said Rāma, 'where then and to what place did you see the *rākshasa* take my wife?'

'Alas!' said Sugrīva, 'I know not where that demon lives, but this I promise you, that I and my dependants shall seek him out. May I ask too that you, who know of my distress, help me against Vālin who has taken from me wife and kingdom?'

'Be assured that I will help you and be at your side from this moment on,' replied Rāma, 'for your sorrow is the same as mine, and friends must come to each other's aid.'

'You may not know,' said Sugrīva, 'how great is the task that lies before you. Vālin is most mighty and skilled in battle. Once, when the Asura Dundubhi in the form of a buffalo challenged the ocean to battle, Vālin fought Dundubhi on behalf of the seas and the mountains and, after terrible battle, he killed Dundubhi and cast his corpse a league away. The body fell close by here; close to here also grow the seven lofty *shāla* trees which Vālin could strip of leaves in a single moment. Such is his strength, noble Rāma, and I fear we may not be his equal.'

'Show me where Dundubhi's corpse lies!' said Rāma.

Thereupon Sugrīva led the way, and when they came upon the dead and withered Dundubhi, Rāma picked up the body and, hurling it mightily, cast it full three leagues away. Seeing then that Sugrīva still appeared in doubt, he raised his bow and with one powerful arrow, cut down all seven *shāla* trees. At once Sugrīva's spirits rose, and thereupon without delay they set out for Kishkindhā where Vālin dwelt.

Alas for Sugrīva! The first encounter left him lucky to escape with his life.

'Why did you not rush in and help me?' Sugrīva complained to Rāma.

'It was because I could not tell you apart,' answered Rāma, 'let us try again, and this time you shall wear a *gajapushpī* flower so that I may know you from your brother.'

Then they all returned to Kishkindhā and, concealing themselves in the trees, they heard Sugrīva let out a fearful yell which pierced the very heavens. At this, Vālin, more enraged than ever, came out from his city to give battle and, when he encountered Sugrīva of the golden hue, a terrible fight arose between them. Then Rāma, from where he stood in the forest, saw Sugrīva weakening and looking about him again and again. At this he fitted an arrow to his bow and loosed it at Vālin's heart. Like a fire which the flames have left, Vālin wilted and sank to the ground. As Rāma came forward then with Lakshmana following him, Vālin turned to him and said:

'You who are noble and the son of a king have slain me thus with an arrow while I fought with another. You, Rāma, renowned throughout the world as merciful and compassionate, knowing what befits every occasion and of firm resolve, how can you show your kingly virtues thus? Though Sugrīva rightly fought me for this kingdom, think well whether your share in the fight was an honourable one.'

At this reproach from Vālin, Rāma replied:

'Not for you to blame me thus. I hold sway over the place where you dwell. You know nothing of what is right and fitting in that you have stolen your brother's wife, Tārā, and taken his kingdom. Know then that I, as upholder of justice in Bharata's realm, have executed you for your crimes.'

Then Vālin saw the nature of his sin and begged Rāma to look after his son Angada. 'May I pray you,' continued Vālin, 'that the fair Tārā be not punished, for the fault was mine in leading her away.' Then, turning to his brother Sugrīva, he said: 'Alas, that for us fate decreed an enmity, alas that the love of brother for brother prevailed not between us. It is for you now to rule over the dwellers in the forest, so take the diadem from my brow. Care especially, I beg you, for my son Angada, my son who lies by me here on the ground, his eyes full of tears, brought to sad wisdom in his tender years. Protect him from harm and carry out Rāma's commands, for not to do so were a sin, and if you spurn his law, he will punish you.'

Then Vālin's head sank back, his mouth opened, showing his fierce teeth, and the valiant monkey died.

Bitter was Tārā's grief and that of Angada, her son. Sugrīva too lamented that he should have lost a brother thus, but Rāma consoled them and told them how none could escape the workings of destiny. Then Lakshmana stepped forward and commanded that a funeral pyre be raised, whereupon the body of Vālin was despatched with all the funeral rites, while the monkeys poured libations of water.

When Sugrīva had been installed as king and Angada as crown prince, the rainy season had begun, so Rāma and Lakshmana dwelt in a cave on Mount Prashravana and awaited the month of Kārttikā, when Sugrīva had promised to begin the search for Sītā. When autumn came and still there was no sign of Sugrīva, Rāma began to doubt his ally's word. Thoughts of Sītā kept his sorrow fresh, and Lakshmana in his attempts to comfort him became ever more angry with the monkey king.

Meanwhile in Kishkindhā, Sugrīva, lost in the delights of his new-found realm and letting the days pass by unnoticed in the company of his wives, was one day warned by Hanuman that the time to help Rāma had long since come. Sugrīva then, mindful of his promise, commanded Nīla to summon the army

and returned forthwith to his merrymaking. Suddenly a great
alarm arose outside the town, and the ministers came hastily to
Sugrīva to announce that Lakshmana had arrived at the gates,.
but Sugrīva heeded them not. Then Angada entered the palace
and declared to the king that he should bid Lakshmana welcome. But Sugrīva, sunk in merrymaking, paid him no attention. Finally Lakshmana, terrible in his rage, began to enter
Kishkindhā, and the noise of the frightened apes as they scattered in every quarter brought Sugrīva to his senses. Hanuman
counselled him to receive Lakshmana with all due humility and
respect as in his folly he had let the appointed time slip by.
When Lakshmana entered the palace, only Tārā's courteous
pleading saved Sugrīva from the Kshatriya's wrath.

'Forgive him that he should have passed his time in love,
heedless of the days. At least he has summoned his array, and
the army now stands prepared to deal with Rāvana's demon
hordes.'

Lakshmana, somewhat calmed, asked that they should forthwith set out to Rāma's abode, whereupon Sugrīva gave orders
that all the monkeys from every quarter of the earth be summoned. Then, as the day waned, they approached the cave on
Mount Prashravana, and Rāma came forth to meet them.
Sugrīva, gazing up at the white-flecked night sky, the moon's
disc pure from stain and the autumnal night anointed with
silver rays, deemed the time propitious.

'My great hosts are now assembled,' said he to Rāma, 'all
obedient to my command; say now, great hero Rāma, what you
would have us do.'

'First I must know whether my Sītā still lives, and where the
*rākshasa* has his dwelling. When we have found this much out,
then we can take further counsel. In this matter, I put my fate
in your hands; take what measures you will.'

Thereupon Sugrīva sent his hordes to all quarters of the
earth; those under Vinata to the east, as far as the mountain of
the rising sun; to the south a troop with Hanuman and
Angada at their head. Then to the west, a great force under the

command of Sushena, Tārā's father, and Shatabala towards the Himālaya he dispatched, to search the northern quarter. To each and every leader Sugrīva gave close details of the regions they would encounter, and when Rāma wondered at so great a knowledge of the world, Sugrīva said to him:

'You remember that I told you of how my brother slew Dundubhi? When Vālin was pursuing Dundubhi, the fearful chase led to every corner of the earth, so in my brother's company I learned the features of every land.'

Ordering the monkey hosts to return with news within one month, Sugrīva especially commissioned Hanuman, the noblest and most able of his commanders, to bring the task to a successful end. Hereupon Rāma came forward and addressed the mighty ape:

'Since you are chosen above all and especially entrusted, I give you this ring with my name upon it, that it may be taken to Sītā when you find her.' The worthy Hanuman bowed and, taking the ring, raised it to his head; then bowing once more at Rāma's feet he took his leave.

Yet, though the monkey hordes searched in every corner of the land, the month crept slowly by and at the last day, when Hanuman and his troops had combed the Vindhya forests, they gathered together and sat in solemn council.

'The time has gone,' they said, 'and if we return with our mission uncompleted, we merit no fate better than death.'

Hanuman was so distressed he was preparing to kill himself when the other monkeys intervened and said they should perform a sacrifice. At that moment a vulture who had witnessed their distress and indecision, swooped down from the skies, thinking that the ape's near death would provide him with a meal. But Angada ran forward to meet him as he approached, and piteously told him of their plight.

'We seek the *rākshasa* whom one of your brethren saw carrying off Sītā. Jatāyu it was who told Rāma that his wife had been snatched away by Rāvana, but we cannot find where Rāvana dwells.'

The vulture then, abandoning his first intent, was led by Angada back to where the monkeys, joined by their allies, the bears, sat in sad array.

'Know me to be Sampāti,' said he in deep and comforting tones, 'the brother of Jatāyu I, and though my wings were burnt away when once, with my brother, I flew too near the sun, and though my strength has therefore gone, I can still be of aid to Rāma, for I too have seen a fair lady, carried through the air by Rāvana in his chariot, and all the while she wept and cried: "Rāma, Rāma, Lakshmana!" Sītā therefore it must have been, and I can tell you too that the *rākshasa* who stole her, that Rāvana, dwells in the city of Lankā. It lies in an island, a hundred leagues over the ocean and was built by Vishvakarman, the artificer of the gods.'

And as he spoke and told the wonderful tale, his wings grew out anew, as the *rishi* Nishākara had promised him they would, if he did but perform a service to the mighty Rāma. But ere his wings had reached full size almost, the excited monkeys flew swiftly to the ocean's shore, eager in the hope of finding Sītā. But when they gazed across the sea, as impossible to cross as is the sky, the monkeys once more sat round in puzzled consultation. Each tried how far he might jump, but none there was who could leap a hundred leagues. Not even Angada in all his noble youth and strength could compass the fearful distance. Then Jāmbavan, king of the bears, arose and spoke to Hanuman:

'You are the son of Vāyu, god of the wind, expert in every science. Why do you sit there disconsolate and alone, son of the wind? Arise and cross the mighty ocean; strike out with your swift speed even as the great god Vishnu encompassed the world with his three steps.'

Then Hanuman, finding fresh confidence and energy born of Jāmbavan's flattering words, swelled out his frame, breathed in the winds and waved his great tail about for very joy. In his new-found strength, the mighty ape arose and forthwith climbed nearby Mahendra mountain, and the mountain groaned and roared like a great elephant assailed by a lion. All was still, for every living being, ascetics as well as creatures of the woods, forsook its tree-clad slopes when Hanuman stood at the peak, alone and splendid in his great endeavour.

## THE EXPLOITS OF HANUMAN

As the great Hanuman prepared himself for the mighty leap that would take him across the hundred leagues of ocean to Lankā, Mount Mahendra trembled and shook to its very base, and the birds wheeled frightened in the sky, crying shrilly as at the approach of thunder. Then, like an arrow from Rāma's bow, Hanuman flew up into the air and high across the waves. The ocean god, Sāgara, stared up in wonder at the tremendous form, shooting over his realm, and, fearing that harm might come to the ape, he bade a reef rise up in the middle of the sea, so that Hanuman could rest awhile. But though Indra praised the lonely rock, and though the rock itself begged Hanuman to take a rest upon his way, reminding him how in ancient times his father Vāyu had helped the mountains when Indra clipped their wings, yet Hanuman swept onwards through the clouds. Monsters rose out of the ocean and threatened to devour him; Surasā, the mother of snakes, and the terrible Simhikā, but Hanuman, changing his size at will, slipped through Surasā's fangs and slew Simhikā.

On, like Garuda the eagle, Hanuman sped, until, reaching the distant shore, he saw before him the city of Lankā, floating in the sky as it were, with its cloud-like white mansions covering the peak of Trikūta mountain. Here was the city built by the divine Vishvakarman, where Rāvana the cruel *rākshasa* held sway. Hanuman despaired of entering it unseen and waited until nightfall before he dared, making himself as small as an ant, approach the gate. Yet even as he entered the city, with its terraced houses, the goddess of Lankā barred his way through the city gate. Appearing as a giant *rākshasī*, she forced Hanuman to battle, but he by his superior might swiftly defeated her and slew her. As she died she said to Hanuman how Brahma once had told her that her death would spell disaster for Lankā.

Then Hanuman pressed onwards into the centre of the town and there he saw, lustrous in the rays of the moon, the palace of Rāvana. Countless were its terraces, its courtyards spacious, and here and there Hanuman could see the hundreds of fair ladies of Rāvana's harem, yet amongst them all no sign of Sītā could he see. Searching through every room of the palace and every court and hall, examining Rāvana's apartments, the rooms where his wives lived, and Pushpakā, the aerial chariot, Hanuman covered every single spot, but still his search remained in vain.

Despondent then, he thought to himself, 'How can I return with empty news like this? What will the other monkeys say to me? What remains for me, now that my quest has failed, but either penance or death?' Then, wandering through the courtyards, he came out into a grove of *ashoka* trees and there he beheld Sītā, like a ray of moonlight filtering through the leaves. Nor was she easy to recognize, for sorrow had emaciated still more her slender body, and the vow of refusing all Rāvana's attention had left her clad only in the garments she was wearing when first he bore her away. These now, worn and tattered, covered her pitiable frame, weak from want of nourishment: no ornaments bedecked her as Hanuman gazed at that tear-stricken face, lost and bound up in its own grief. Hanuman reasoned with himself, making sure by the various signs he had been told by Rāma that it was indeed Sītā, and then, as he was about to approach her, he heard the sound of music and, turning round, beheld Rāvana drawing near. Struck with wonder at the terrible grandeur of the *rākshasa*, coming onward surrounded by maidens bearing torches aloft, Hanuman crept swiftly in among the trees of the *ashoka* grove.

As Rāvana then came up to the dark-tressed Sītā, his gestures spoke of love, and Hanuman could hear him saying to her:

'Sītā, with your large eyes and fair face, you have stolen my heart, please become my wife. It is folly to continue thus pining for Rāma: marry me and become the foremost of my wives. What use is Rāma to you now, an exile and an ascetic? You know not even whether he still lives.'

But Sītā threw a straw between them and answered him, 'Think not of me, but rather regard your own people. I am the wife of another and therefore you must not look upon me; this is the law of the righteous. My husband is merciful to those who seek his compassion: if you wish to live therefore, make your friendship with him. Restrain yourself, thus it behoves you, and take me back to Rāma; if you do this, good fortune will be yours, but if you do not, O Rāvana, you will surely go to your death.'

When the king of the *rākshasas* heard these stern words, he answered fair Sītā harshly: 'Strange indeed are the ways of love, that the object of it is protected by affection. Only for this, Sītā, I do not kill you. Of the limit that remains to you, two months more will see its passing; then, if you will not be mine, I shall have you thrown to the kitchen and prepared for my morning meal.' With this the angry *rākshasa* turned and withdrew the way he had come, making the trees shake and the earth tremble with the madness of his wrath.

Hanuman then, having learnt the truth of the matter, determined to speak gentle words of Rāma to her and, swinging lightly from tree to tree, the son of Vāyu came to where she sat, bowing low, with his face of coral red. He placed his hands upon his head and forthwith addressed her:

'I believe you are Sītā, dear lady, and surely the queen of Rāma.'

With delight at the mention of her husband's name, Sītā spoke to Hanuman in his place upon a branch: 'I am indeed the daughter-in-law of Dasharatha and daughter of Janaka, king of Videha. Sītā is my name and I am the wife of noble Rāma.'

Sītā told Hanuman her story while the noble monkey listened, and when she had finished, he said to her:

'I come, my queen, at the command of Rāma; he is well and ever anxious for your welfare.'

Hanuman related how they had all searched for her after Rāma had made alliance with Sugrīva and how they had found the ornaments which she had let fall when Rāvana carried her off.

'Now take this ring,' he concluded, 'which is marked with Rāma's name and was given to me by the hero so that you might have confidence when I saw you.' As Sītā took her husband's ring and looked at it, a joy filled her heart as if she had actually regained her husband.

'Tell me, Hanuman, best of monkeys,' she asked him then, 'is Rāma truly alive and well? Why does he not come and burn these demons with his wrath? But they are mighty, I suppose, and it will be time before he can lay them low. But has he made no plan, taken no steps, to come and rescue me?'

With greatest respect, Hanuman gently replied:

'Fair lady, Rāma does not even know where you are. As soon as he hears from me he will come here without delay, bringing a mighty host of monkeys and bears; with his deadly arrows ridding Lankā of every *rākshasa*. Your name is ever on his lips; worried with the sorrow of not seeing you, he sleeps but little, and that without peace, nor does he take interest in food. I pray you now, if you will not come with me yourself, give me some token, that he may recognize it, and give me too a message that I may bear to Rāma and Lakshmana.'

'Tell Kausalyā's son that I ask after his welfare,' she replied, 'and that he remains ever my lord. Tell Lakshmana too, him of few words but most dear to my lord's heart, that I ask for him. Now, Hanuman, none but Rāma may rescue me, so I cannot come with you, and since you are my only messenger, give Rāma this message: "I shall live for yet one month, Rāma, longer I cannot live; this I swear to you."'

Thereupon she took a corner of her garment and, untying it, gave Hanuman her radiant crest-jewel, sparkling in the night air. 'Give this to Rāma too,' she said.

Hanuman bowed, taking the jewel, and, making a reassuring farewell speech, withdrew from Sītā. But the great monkey reflected that there were still some tasks for him to perform before he left the island of Lankā. Thinking that to have carried out his main errand was not sufficient, for only those of niggardly mind do exactly what is allotted to them and not one whit more, Hanuman decided to destroy the *rākshasa's* pleasure-grove and test the strength of the enemy. So, swelling his form to gigantic size and terrifying the *rākshasīs*, he tore up shrubs and trees and left the grove as desolate as if a fire had swept it. The fleeing *rākshasīs* told Rāvana: 'Your majesty, a fearful ape has destroyed the whole of the *ashoka* grove, except for that place where Sītā is; we know not why.'

When Rāvana heard this, he forthwith commanded eighty thousand *kinkaras*, the servants of the demons, to catch the savage ape. Powerful though they were, their numbers helped them not, for Hanuman, seizing a mighty club, swung it on their skulls as an unkind wind bends low the ripening corn. The few that remained fled in terror to tell their king of the disaster. Meanwhile Hanuman tore down the *rākshasa's* temple and as the temple guards ran up, he set fire to what remained and slew them all.

Then Jambumālin, son of Rāvana's minister Prahasta, set forth at Rāvana's command, but him also the worthy monkey slew with no great pains. There followed then to the slaughter seven more ministers' sons with their companies, but Hanuman routed them utterly. Then five chiefs of the army came out to

settle with the terrible intruder, but the great ape crushed the
first to death, killed the next two with an uprooted tree and
slew the last two by snatching off the top of a mountain and
hurling it at them. No less a warrior then than Aksha, Rāvana's
own son, set out to catch the ape and suffered a swift death. In
desperation the demon king entrusted his other son, Indrajit,
with the capturing of Hanuman.

A terrible battle ensued, a wonder to all who beheld it.
Indrajit was versed in every art of warfare, and the fight raged·
until Indrajit let loose an arrow given to his father once by
Brahma. The monkey then, bound hopelessly by that powerful
weapon, sank down upon the ground. Tied like a maddened
elephant he lay there until led before the *rākshasa* king. Though
tortured on every side by the demons thirsty for revenge, Hanuman could not help but admire the magnificence of the demon
king whom rage, it seemed, made yet more radiant. Then, looking at him and maintaining a noble bearing, Hanuman spoke
to him, being asked why he had wrought such destruction.

'I am the son of the wind god, Vāyu,' he said, 'Hanuman is
my name. For Sītā's sake I have crossed the ocean of a hundred
leagues, and Sītā I have found here, kept prisoner by you. Know
that I am the messenger of Rāma, come to see you, but that I
have had to fight my way to your presence. The message is this:
you, who know what righteousness is, should give Sītā back to
her lawful lord. These are Rāma's wishes, and none disobeys
them without regret.'

When Rāvana heard these haughty words his fury broke out
like a thunderbolt from the sky. 'Kill him, take him away and
kill him!' he roared. But his brother, Vibhīshana, standing at
his side and ever wise in counsel, said:

'One come as a messenger must not be killed; such is the
inviolable law between enemies.'

'It is true what you say,' replied the *rākshasa* king, 'then let
it be like this: since monkeys prize their tails, set light to his and
let him go.' At this command the demons wrapped torn shreds
of cotton to Hanuman's tail and poured oil on them.

Meanwhile in the desolation of the *ashoka* grove, her guardians had kept Sītā informed of the terrible events. When Sītā heard that Hanuman had been captured, she felt all hope had gone, but Vibhīshana's daughter came hurrying with the news that Hanuman was to be set free, though the angry demons were setting his tail alight. At this the gentle Sītā began to pray to Agni, god of fire, asking him that he might not use his power on Hanuman, but leave him unharmed. Agni heard her prayer, and though the *rākshasas* rejoiced as flames leapt up from the monkey's tail, Hanuman felt no hurt nor was a hair on his body singed.

Escaping from his tormentors, Hanuman withdrew outside the city and, looking down on Lankā, considered the result of his deeds. I have destroyed the *ashoka* grove, thought he, brought down the temple, destroyed part of the army and killed the cream of the *rākshasas*. Now what remains? The fortress itself! This I should manage with my tail; I shall make a fire sacrifice in Lankā.

Coming to this decision, the great monkey leaped up on to the terraces of the houses and, curling his fiery tail this way and that, he set every one alight. The dry wood flared up till the whole city lay ablaze. Then Hanuman ran down to the sea and, dipping his tail in the waves, put out the flames. Making ready to return, he sprang first to where Sītā was and, seeing that she was safe, he climbed a mountain to prepare for his leap across the ocean. Once more he made the mighty flight, his feet striking so hard upon the peak as he took off along the sky that it was levelled to the plain, and the mountain sank to the nether world.

Seeing him land on Mahendra mountain, all the monkeys gathered with cries and shouts of great delight, bringing the unscathed hero offerings of the choicest fruits. Then Hanuman, bowing to all the elders among the monkeys and to Jāmbavan, king of the bears, began the tale of his great exploits, not denying himself a bit of boasting here and there.

Angada counselled that the most powerful among the mon-

keys should bring Sītā back, but Hanuman told them that this
was not suitable, for he himself had offered to carry her over
the ocean on his back, but Sītā had refused; only Rāma should
touch her, only Rāma rescue her. Jāmbavan also disapproved
the plan and said their duty lay in returning forthwith to Rāma.
At this command the monkeys made eagerly for Mount Pra-
shrayana, stopping on their way in the bee-filled glades known
as Madhuvana.

There, as guests of Dadhimukha, Sugrīva's uncle, they cele-
brated the success of Hanuman's deeds, but alas! somewhat too
well. Drunk on the glorious honey, the monkey hordes swung
tipsily from tree to tree, breaking branches, pulling out shrubs
and causing terrible destruction. The guardians of the glades
fled in terror, and poor Dadhimukha, rudely repulsed by the
drunken Angada, gave up trying to control his troublesome
relations and went forthwith to his nephew.

There, before Sugrīva, Dadhimukha related the sad story of
the destroying of Madhuvana. When Lakshmana heard that a
monkey had visited Sugrīva with news, he thereupon approached and asked why Dadhimukha had come.

'No news of Sītā does he bring, alas!' exclaimed Sugrīva,
'only the ruin of his precious bees. Yet I am sure that this wild
celebration of our tribe which has caused my uncle's grief bodes
good for the outcome of their task.' Then, turning to Dadhimukha, Sugrīva continued, 'Return quickly and bring the
monkey army back here with all possible speed!'

The disconsolate Dadhimukha made his way back and told
Angada of Sugrīva's command. Angada, by this time somewhat
sober, speedily agreed to obey, and so the whole great host sped
through the air in great jumps and leaps till they reached Prashravana mountain.

From the vast assembled army, Hanuman stepped forward and, bowing deep before Rāma, spoke these words to
him:

'On an island called Lankā, a hundred leagues from the shore
of the southern ocean, lives Rāvana, king of the *rākshasas*.

There, in a grove of *ashoka* trees, part of his palace, I found Sītā, alive and thinking only of you her lord. I comforted her and told her of your friendship with Sugrīva, whereat she was greatly delighted. I gave her your ring and in return she made me take this crest-jewel. She sent messages of goodwill to you and Lakshmana and asked me to repeat these words to you: "I shall live for yet one month, Rāma, longer I cannot live; this I swear to you." Such was her message and here is the jewel.'

When Rāma heard these words and gazed at the beautiful jewel, recalling to his mind yet more vividly the face of his beloved, he could not contain his grief.

'She talks of a month, but I cannot live an instant longer if I do not see her,' he lamented. 'Take me to the place where she is, I cannot wait. Tell me what did my beloved say to you, with her sweet voice?'

Hanuman, moved to compassion by the noble hero's sorrow, said:

'When she had asked after your welfare and that of your worthy brother, as well as of Sugrīva and his ministers, she begged me that you all should do what lay within your power to rescue her from that dread demon king. And I, to comfort her in her great distress, spoke these words of farewell to her: "Those two lionhearts, Rāma and Lakshmana, will soon appear, mounted on my back, as though the sun and the moon had risen in the sky before you. Fear not but you will see, at the very gates of Lankā, the destroyer of evil-doers come to rescue you. For I can tell you that you will yet sit by Rāma's side in Ayodhyā when the years of exile have passed." These words I spoke to her, most noble Rāma, and she took courage.'

## RĀMA'S CONQUEST OF RĀVANA

When Rāma heard Hanuman's words, he took them to his heart in supreme delight and then, turning to Sugrīva, said:

'Let us move off this very hour, nor should further time be lost. My one great fear remains though, worthy Sugrīva, and it is this: how shall we cross the ocean?'

Sugrīva answered him and said: 'Have no fear on this account, for we shall build a bridge across to Lankā. Hanuman can tell us all about the strength of the enemy, and so I can foresee successful outcome for our venture.'

Thereupon Sugrīva summoned the mighty host to prepare to march, and Rāma himself gave the order to set forth. The horde of monkeys and bears, covering the whole earth, it seemed, drew on over Sahya and Malaya mountains.

When, at evening, they reached the shores of the ocean, Rāma looked at his brother who stood at his side and said to him:

'My sadness should be on the wane, but somehow it grows greater every moment that I am kept from Sītā. Feel this breeze that blows in from the waves, perhaps it has caressed her; and the moon rising, look at it, perhaps she is looking at it even now.'

Then Rāma went to offer his twilight prayers.

Meanwhile in the city of Lankā, Rāvana called an assembly

of the *rākshasas* and, somewhat crestfallen when he surveyed the fearful destruction which Hanuman had caused, he spoke to the attendant throng:

'This city which I thought could never be conquered, has been devastated by a mere monkey, and Sītā has been discovered. I ask you now to think well upon what this betides and seek what means you can to prevent Rāma from entering here, for assuredly he will come, nor will the ocean prove an obstacle to him.'

Then with various voices the resplendent *rākshasas*, raising their jewelled swords and shining with an inborn light, praised Rāvana and extolled his former mighty deeds. When they had thus restored their leader's confidence, many boastful words were heard.

'We shall kill Rāma as soon as he sets foot on Lankā, as well as his brother Lakshmana and Sugrīva with his monkey hordes. All shall be laid low, not even the wretched Hanuman who caused us so much woe shall live to see the other shore again.'

But Vibhīshana, Rāvana's younger brother, listened to all this boasting with growing discontent.

'Put down your bright swords,' said he suddenly, 'sit around me here upon the ground and hearken to what I say. You know well that Rāma has never offended our king in any manner; to use force therefore to gain and keep possession of his wife is a wrong and sinful act. To steal another man's wife is to destroy one's good name and one's life; you may be sure then that the daughter of Janaka kept here in bondage will bring us no good fortune.'

At these words the assembly dispersed, and Vibhīshana, thinking that he had somewhat gained his audience, sought out his brother the next morning.

'The omens are against us,' said he, 'I see the signs that threaten our destruction ever on the increase since you took Sītā from her husband.'

'Let us once more summon the assembly,' said Rāvana, angered that his brother should seek to thwart his designs. Then

Rāvana, resplendent in majestic garb, radiant with his jewelled ornaments, strode purposefully to the council chamber.

'Prahasta, summon the guard and bid them be constantly prepared against attack!' he commanded, whereupon Prahasta left the hall and went to make the city safe from sudden and surprise defeat. Present in the assembly was the giant Kumbhakarna, awoken from his six months' sleep, and when Rāvana asked him what should be done, he merely laughed and rebuked him for making so much trouble about a task which was as good as done.

'I promise you,' said Kumbhakarna, 'my strong arms in that battle about to come; have no fear but that Rāma and his brother assuredly will die.'

'I have no doubt you are right,' said Rāvana. 'Surely Rāma cannot know my strength, else he would not dare to come.'

Once more then, Vibhīshana's sage counsel could be heard.

'It is not right for us to make this useless quarrel with the noble Rāma who is righteous and who has done us no harm. You should give Sītā back to him.'

'We have no fear of anyone, Rāma or the rest,' Prahasta broke in.

'Take care,' answered Vibhīshana, quiet and insistent, 'for Rāma is a mighty warrior who has already shown his strength in battle; I warn you not to follow the fate of Khara, who lost his life trying to avenge Shūrpanakhā.'

At these words of his uncle, Indrajit spoke out insultingly:

'I did not think to hear my father's brother speak such weak and cowardly words.'

Whereupon Vibhīshana turned to him and answered:

'You are but a foolish child, nor should you have been allowed to appear in this assembly.'

Meanwhile Rāvana, listening with mounting ire to his brother's advice, let fall his wrath upon the noble Vibhīshana.

'You are like a snake within my house; worse, for you talk like a friend but your heart is with the enemy. If anyone but you had dared to speak thus in my presence, I would most certainly have slain him. Shame on you, for a disgrace to the house of Rāvana!'

At these fierce words, Vibhīshana, holding his mace, flew up into the sky and, accompanied by four other *rākshasas*, made a parting speech to Rāvana.

'Though you are my brother and may say what you like to me, I tell you, even though I must respect my older brother as a father, that you follow a wrong path. Those who speak what is displeasing for another's good are few, but rarer still those who will listen. Now I go upon my way.'

With this, Vibhīshana flew off and reached the ocean's other shore. The monkey host, on seeing the five *rākshasas* alight in their midst, seized their weapons and prepared to slay them, but Vibhīshana told them he had come with peaceful purpose and, stepping forward to where Sugrīva and the monkey leaders stood, he at once announced himself:

'I am Vibhīshana, younger brother of Rāvana, king of Lankā, whose unworthy deeds are known all too well to you. He who carried Sītā away and killed Jatāyu has been warned by me to restore the wife he stole and make his peace with Rāma. But some strange fate must drive him on, for he has not heeded my advice but only spoken to me harshly.'

When Sugrīva heard this, he turned to Rāma, saying:

'I trust not these *rākshasas*. What if Vibhīshana has been sent by Rāvana to spy out our purposes? My advice is that we should punish him.'

But Hanuman, who knew of old Vibhīshana's forbearance and how he once had saved his life, came forward then and said:

'You may trust Vibhīshana, for he has seen far into his brother's evil mind. Your valour, noble Rāma, he respects, of this I am sure, and I do not doubt he will fill Rāvana's place more worthily when we have settled with that *rākshasa*.'

When Rāma heard this well considered speech from Hanuman, he gave as his solemn judgement:

'One who has come as a friend should not be forsaken, even if there be some evil intent in him. He can do me no harm, and

since he has sought my protection I must honour my vow to grant safety to all creatures.'

Vibhīshana then bowed humbly before Rāma, saying:

'I have left my homeland, wife, children and possessions, all these I have forsaken; to regain them now, I am entirely dependent upon you.'

Thereupon Rāma embraced him and commanded Lakshmana forthwith that he be consecrated with water from the ocean. Rāma then asked Vibhīshana how they might cross the desolate watery leagues.

'It is my belief,' answered Vibhīshana, 'that if you prostrate yourself before the god of the ocean, he will show himself favourable towards you, and a way across will thus be found.'

While Rāma went down to the shore of the sea, there appeared a *rākshasa*, Shuka by name, in the sky above the monkey host. Sent by Rāvana to make discord in Rāma's following, Shuka seized his chance and swept down while Rāma remained in worship before the ocean. When Sugrīva learnt of his mission he was sorely tempted to destroy him, but instead replied:

'Return and tell your evil master that I will have none of his foul tricks. I have sworn allegiance to the noble Rāma and to him alone I remain faithful.'

Meanwhile Angada commanded his troops to seize the wretched Shuka and bind him helplessly. When Rāma was told he ordered that the *rākshasa* should be freed from his bonds but kept in captivity.

For three whole days then Rāma remained by the shore, passing the night in prayer, but still the ocean god refused to appear. Then Rāma became angry, and his eyes, red from their vigil, became yet more red. Summoning his brother he commanded him:

'Go, Lakshmana, and bring me my bow with its deadly shafts. I shall dry the ocean up, so let us see what Sāgara will do.'

As Rāma stretched the terrible bow and loosed the arrows into the waves, the ocean boiled and seethed in frenzy. Then, from a parting in the mighty billows, Sāgara, god of the ocean, appeared, holding Rāma's arrows aloft.

'I shall do as you will,' said Sāgara. 'Call forth Nala from among the monkey hordes; he is the son of Vishvakarman, and when I have made the waters still, that son of the divine architect will build you a bridge to Lankā.'

Thereupon Sāgara sank beneath the waves, and the waters grew still and abated over the hundred leagues. When Nala had fashioned the bridge, the army of the king of monkeys crossed swiftly over and took up positions on Lankā's shore, first provisioning themselves with the abundant roots, sweet water and delicious fruits which they found there. Then Rāma, surveying the vast array, commanded them to proceed towards the city. The sky became turbulent with the yells of the excited monkeys, and from the city could be heard the shrieks of the *rākshasas*, whereat the monkeys roared the more. Rāma gave orders for Shuka to be released so that he might herald his coming to Rāvana. Shuka, flying swiftly through the air, entered the royal palace and announced himself to the *rākshasa* king. When

Rāvana heard of his adventures, he could not believe the report.

'Go back!' said he, 'take Sharana with you and let the two of you bring me some sober news. I will not believe all this excited nonsense.'

Then, taking Sharana with him, Shuka once more flew over the monkey host, but in their eagerness to spy out the monkeys' strength, they were caught by Vibhīshana. Once again Rāma let them go free, and on their return to Rāvana's presence, Sharana spoke to him:

'All that Shuka told you is true,' he said, 'Rāma marches towards the city with a vast army of monkeys and bears. Let us go to the palace roof and I will point them out to you.'

Unbelieving still, Rāvana mounted to the topmost terrace, and there, spread out before his city, like an endless autumnal forest, lay the unnumbered strength of Rāma's vast command. Excitedly the two messengers, Shuka and Sharana, pointed out a chieftain here, a leader there, and gave Rāvana their names.

'Enough of this,' said the demon king. 'It seems to me that you are lost in admiration. Away with you!' he shouted angrily, still not wishing to believe; but when a third *rākshasa*, the powerful Shārdūla, confirmed all that had been said, Rāvana gave orders for his council to be summoned. So great, though, was his rage that no advice could please him. Thinking once more of the cause of all the turmoil, he commanded Vidyujjihvā to conjure up by means of his magic powers, Rāma's head and Rāma's bow. This done, he set out to the *ashoka* grove and, showing Sītā the false head and bow, he said to her:

'Your husband and his army have been defeated in battle overnight. Here is his head and here his bow, captured in the fight. What use, fair lady, to regret him more or weep for him? Say only that you will have me as your lord.'

When Sītā heard these terrible words like the arrows of death, she fainted away, but then, regaining her senses, she wept and lamented. 'Never will I be yours, foul wretch!' she said. 'Now that you have killed my husband, I seek only death. Go, leave me here to die!'

When Rāvana saw that her heart would ever remain as adamant towards him, the angry *rākshasa* went off and, summoning his ministers, commanded them to make the army ready for action.

In the meantime Saramā, Vibhīshana's wife, who had observed the demon king's vile trick, came out to the *ashoka* grove and consoled the sorrowful Sītā.

'It was but a device,' said she, 'to make you abandon your lord. Thus could the evil king have averted war, so he hoped, and told Rāma that you had forsaken him in order to become another's wife.'

'Is there nothing we can do to avoid this terrible war?' Sītā then asked Saramā. 'Go quickly to the assembly and find if they have succeeded in making Rāvana return me to my lord.'

The worthy Saramā departed and ere long returned.

'Alas, dear Sītā, the news is bad. Though Mālyavan advised that you should be given back, that Rāvana was completely lost in sinful action and that all omens pointed towards a fearful fate in store for them, yet the king's heart was hardened and in his rage he has ordered that the city gates be manned.'

Meanwhile from Rāma's camp Vibhīshana had sent out his four followers in the form of birds to spy out Rāvana's preparations for battle. Rāma then decided to survey the city from Suvela mountain and so, accompanied by his brother, with Vibhīshana and Sugrīva, he climbed up to the lofty peak. There, looking over the city, they saw the *rākshasas* frantically building a wall to protect themselves. On the mountain-top there, they passed the night, and when morning came, they saw Rāvana looking out from the top of the city gate. Descending then from Suvela, Rāma ordered his army to take up positions to besiege the city and, summoning Angada, he said to him:

'Go, worthy monkey, and bear this message from me to Rāvana, the ten-headed demon: "Wretched *rākshasa*, I bring the rod of justice and wait outside your gate. If you do not restore my wife, I shall sweep the world of *rākshasas* by my own sharp arrows. Vibhīshana I shall make to rule over your kingdom. Foolish and sinful as you are, your time is short.'''

Thus commanded by the noble Rāma, Angada forthwith fulfilled his mission after announcing himself to Rāvana in proper manner. The *rākshasas* were incensed at the proud words, so they laid hands on Angada who had difficulty in making his escape. Though this treatment left Rāma in but little doubt of his enemy's intent, Sushena, Tārā's father, went round the city to inspect the gates and reported that all remained fast closed.

Angered beyond control at the tone of Rāma's message, Rāvana ordered that a force should set out, and as night came on there flared up here and there combat between *rākshasa* and monkey as the foes came to grips. Then Rāma, seeing where the battle lay thickest, came up to help and put the *rākshasas* to flight. Angada at his side overcame Indrajit after a fierce struggle, but the wily Indrajit made himself invisible and, escaping some way off, let loose at Rāma and Lakshmana the self-same weapon with which he once had bound Hanuman. Helpless, Rāma and Lakshmana could not move, whereupon Rāma despatched ten monkeys to recapture Rāvana's son, but Indrajit remained invisible and kept showering arrows on the two heroes. Covered with countless wounds, Rāma and Lakshmana sank to the ground, and while the monkeys lamented, up came Vibhīshana who had the demon power of being able to see his nephew, and he let fly an arrow at him, wounding him. At this the *rākshasa* withdrew and returned boasting to his father that he had slain his two great enemies. Straightway Rāvana ordered his wife Trijatā to take Sītā in the aerial chariot and show her the dead body of her husband. Flying over the spot in the Pushpakā, there on the ground, surrounded by a sorrowing guard of monkeys, Sītā saw her husband and Lakshmana senseless and bleeding. She could not believe that all the prophecies had been thus belied, nor that she saw her Rāma dead. Her grief would have grown beyond measure had not Trijatā comforted her.

'Rāma is certainly not dead,' she assured her, 'but only

unconscious from his wounds.' So they returned to the *ashoka* grove.

Then, as the apes watched and waited, Rāma slowly regained consciousness and called out to Lakshmana. Sushena came up, that wise and resourceful monkey, and told of two wonderful healing herbs to be found on the mountains Chandra and Drona, and that Hanuman should be sent to fetch them. Suddenly, before the task could be carried out, the monkeys saw the eagle Garuda come through the skies, flaming like fire. Garuda flew down and spoke to Rāma:

'Worthy and most noble Rāma, of most bright fame, I am your friend, and your very life, though I move about outside you. I have flown here on my wings to save you both and make you well. The *rākshasas* are ever deceitful in battle, so never trust them to fight as other men.'

With this, the radiant bird embraced them, made them whole and then flew off as a lightning flash across the night sky. Loud and noisy was the monkeys' joy, and the yells of their delight reached the ears of Rāvana as he sat in council with his ministers.

'This loud jubilation bodes no good, but rather gives me alarm,' said he. 'Send out another force.'

Thereupon the fearful Dhūmrāksha sallied forth from the western gate, but not even he could prove a match for Hanuman who picked up a mighty boulder and, shattering first his chariot, then slew the *rākshasa* himself. Then Rāvana sent forth Vajradamshtra who sped out from the southern gate, but he and his forces met swift and bloody deaths at the hands of Angada and his monkey host. Close at his heels followed Akampana, recklessly wading into the dusty turmoil of his fellow-demon's fate. There, though the monkeys at first recoiled from this unexpected onslaught, uncertain in the clouds of dust which was friend or foe, Hanuman soon came up and, engaging in single combat with Akampana, laid the demon low.

The next morning Rāvana reviewed his various troops and learnt of how the battles had proceeded. He therefore sent Prahasta, with four ministers, and he, despite the evil omens, led his forces out through the eastern gate. Yet another fearful battle ensued, though it was not long before the four proud *rākshasa* ministers lay silent in the dust. Prahasta nonetheless fought bravely on, pressing the monkey besiegers back an arrow length here, a dagger length there. To his misfortune though, Nīla came to the rescue of his troops and, challenging Prahasta to battle with club and mace, soon sent the demon to join his fellow ministers. Seeing their leader fallen, the remaining *rākshasas* withdrew hastily to the city.

Beside himself with fury at the constant repulse of his armies and at the failure of their sorties, Rāvana himself came out to battle at the head of a mighty host. The foremost of the monkeys sallied forth to turn him back, but not all their strength availed them against such a ferocious foe. As Lakshmana saw the chiefs of the monkeys being routed and driven headlong before the approaching demon king, he turned to his brother and said:

'Permit me, dear Rāma, to go out and settle with this demon; if he is not soon repulsed he will overrun our camp. Have no doubt, I will soon turn him in his tracks.'

At this bold speech of Lakshmana, Rāma could not but give his assent, whereat his fiery brother rushed out to bar the demon king's attack. Lakshmana let fly his arrows at Rāvana, but Rāvana, brushing them harmlessly aside, kept coming onwards till the two fought hand to hand. A lucky thrust from the demon's lance wounded brave Lakshmana, but at the same moment the worthy brother of Rāma struck Rāvana such a tremendous blow with his fist that the demon fell senseless, stretched upon the ground. Hanuman and Rāma ran forward. Hanuman picked Lakshmana up while Rāma quickly disarmed the demon king. As they looked, they saw that Lakshmana's wound was but a slight one, but meanwhile Rāvana regained his senses and, finding himself unarmed, hastened back to the city.

Once more inside Lankā, the ten-headed Rāvana, reflected on what had taken place and, knowing that no mean heroes would overcome the brothers, he gave the command that Kumbhakarna should be awakened. Kumbhakarna, cursed by Prajāpati to sleep for six months and for six months to remain awake, thus passed the years, but now, alas, was in the middle of deepest slumber. The ingenious *rākshasas*, those eaters of flesh and gore, left no device untried, and finally, to the crash of drums and the pricking of lances, the sound sleeper awoke. Yawning with his cavernous mouth, Kumbhakarna saluted his brother Rāvana and said:

'What is the task you want me to do?'

Delighted at his willingness to serve him, Rāvana embraced him and, his eyes cast low, told him of the misfortunes which had come upon Lankā.

'Since you have been asleep,' he said, 'you do not know what trouble and terror Rāma has made for me. Crossing the ocean with Sugrīva and his monkey hordes, he now strikes at our very roots. All our heroes are slain, only boys and old men are left; so now, my most mighty brother, it is for you to save the city of Lankā.'

When Kumbhakarna heard Rāvana's lament, laughter fell from his mouth and rolled down his chest, shaking his great stomach.

'You would not heed what I said,' he answered, 'and now you are reaping the fruits of what we have foreseen. Do as Vibhīshana said; if not, do what you like.'

Then, while Kumbhakarna yawned heavily again and his eyelids began to droop, like awnings let down against the mid-day sun, the fire of the ten-necked Rāvana's rage broke out:

'Are you my venerable teacher, that you tell me what to do? What use are words now? If you are indeed my brother, you will come to my aid.'

Realizing the state of terror and anger which had caused his brother's bold and bitter words, Kumbhakarna answered him slowly and soothingly:

'Let us have no more of this ranting. Rejoice at Sītā's grief when she sees me bringing Rāma's head from the field of battle.'

At these cruel and boastful words all the *rākshasas* raised a triumphant shout which was heard throughout the island. Then, cavorting over the city wall, the mountainous Kumbhakarna lurched without delay towards the monkey hosts. All the monkeys fled in terror before the vast figure bearing down upon them, but Angada rallied them and brought them back to the attack. In vain was his attempt, for scores fell shattered under the swinging of Kumbhakarna's terrible mace. Then the chiefs among the monkey army grappled with the *rākshasa* but all were routed. Finally Sugrīva rushed forward to the rescue of his tribe, yet he too was knocked unconscious by the gigantic demon and carried triumphantly back to the city. The riotous shouts of the *rākshasas* brought Sugrīva to his senses and, slipping from Kumbhakarna's grasp, he sliced off his nose and ears before making his escape. Maddened with pain and rage, Kumbhakarna flew out of the city again and wherever his mace fell, a band of monkeys lay still in the dust. Then Rāma hurried to the scene and, working his bow, rained arrows upon the giant, who slowly sank to the ground. Rāma, stepping forward, took away Kumbhakarna's mace and ordered the monkeys to make an end of him, but though they swarmed all over him, the mighty Kumbhakarna flung them this way and that, like a vast awning in a gale, slapping and breaking its ropes. Once more Rāma came forward with his mighty bow and shot away the giant's arms and legs, and finally his head.

When Rāvana heard the report of Kumbhakarna's death his lamenting knew no bounds, but still his heart was hard, and, to console him, Trishiras said he would set forth. Then Trishiras, at the head of a mighty force, including Atikāya, one of Rāvana's sons, began to clash with Sugrīva's legions. More bitter than any that had passed was the fighting now, but Sugrīva and Angada, together with Hanuman, in the end proved more valiant than their foes. Only Atikāya was left, but him Lakshmana slew, using the bolt which Brahma had given.

To Rāvana but little comfort now remained, but that little was enough. Still bitterly determined though he knew he could not rout his enemy, nonetheless he toured the city walls and gave stern command that no enemy should enter unscathed. Seeing his father thus vexed and sullen, Indrajit, his one remaining son, promised him he would settle his enemies this time or never. So once again the great warrior Indrajit, skilled in every demon art of war, flew out in his chariot from the city and showered his deadly arrows over the monkey army. Sugrīva's casualties were fearful to behold; like waves breaking on a storm-driven sea, the monkeys ran but a short way and then were toppled over. Not even their chiefs escaped hurt, and before long, the terrible *rākshasa*, making himself invisible as before, assailed Rāma and Lakshmana. Returning joyful to his father, he told him of his exploits, but Rāvana was displeased to know that Rāma and Lakshmana still remained alive.

During that night Vibhīshana and Hanuman went round the battlefield, to succour the fearful heaps of wounded and the dying. Jāmbavan, too, lay there, and when Hanuman approached him, he said to him:

'On Mount Kailāsa, in the Himālaya, four healing herbs are known to me. Go bring them speedily.'

Hanuman flew off, but though he searched Kailāsa, he could not find which were the herbs. At last in his despair he seized the whole top of the mountain and bore it swiftly back to Lankā. Then Jāmbavan drew out the herbs, and with the scent that rose from them, all the great numbers of wounded were once again made whole. Rāma and Lakshmana too, recovered from their injuries.

'Now let no more time be lost,' Sugrīva said, 'for the enemy is weak; fearful he watches us over the city walls; now that it is night, let us creep up and set fire to Lankā.'

The monkey host moved up and, casting firebrands over the battlements, soon set the whole of Lankā ablaze. Then by the city gates terrible encounters broke out, but Sugrīva's heroic hosts conquered the few remaining demon warriors who hurried out to stem their flood. Kumbhakarna's sons fell by the walls of Lankā, one with his neck turned right round in Hanuman's fearful grip.

To Rāvana now but one hope remained. Calling his son Indrajit, he said to him:

'Rāma is mighty, and mighty the strength of his arms, for all the *rākshasas* have now been slain. My heroic son, go forth and kill those two brothers, Rāma and Lakshmana, for whether you fight seen or unseen, you will surely conquer them.'

Indrajit forthwith set out and, wounding many with his arrows, including Rāma and Lakshmana, returned and came forth once more, from the western gate, bringing in his chariot a form of Sītā which he had conjured up by magic. Flaunting the figure of Sītā before the eyes of Lakshmana and Hanuman, he cut off its head. Hanuman, in terrible rage, threw mighty crags at Indrajit, but the latter escaped once more unharmed inside the city. When Hanuman, thinking he had witnessed Sītā's death, told Rāma of what had happened, Rāma collapsed to the ground. Lakshmana then ran to his brother and raised him up, comforting him, and swearing by all the gods that he would make an end of Indrajit.

When Vibhīshana heard Lakshmana's powerful words, he came up and, learning the cause of Rāma's distress, assured him it was but a demon trick.

'Indrajit can perform these things,' he said, 'and now he is offering a sacrifice in the Nikumbhilā grove in order to become invincible in battle; let us attack there and surprise him.'

'Go as Vibhīshana says,' said Rāma then to Lakshmana, 'and since he knows the ground, you will surely kill the demon prince.'

So Lakshmana went to the Nikumbhilā grove and there rained arrows down on Indrajit. A fearful combat ensued between the two in which each lost their weapons. Meanwhile the monkeys and bears closed with the other demons, and as the tumult of battle rose, Lakshmana slew Indrajit's charioteer, and the monkeys felled his horses. Swift as a flash in the confusion of the fight, Indrajit flew back to the city and drew out in another chariot, but this fared him no better than before, for a well-aimed arrow from Rāma pierced the breast of the charioteer, and the horses, plunging wildly, were caught by Vibhīshana. Once more Lakshmana and Indrajit met face to face in mortal fight. Then Lakshmana fixed an arrow given by Indra to his bow and, calling upon the righteousness of Rāma, he drew the feathered shaft back to his ear and let it fly with a thunderous noise. The arrow drew Indrajit's head, with its helmet, out of his body and dashed it to the ground. Seeing their leader's bejewelled head lying bloody in the dust, the *rākshasas* flew fearfully back to the city. Joyful at the outcome, Rāma said to his brother:

'A mighty deed was this, my brother Lakshmana; now that Indrajit is slain, our victory is as good as won.'

In Lankā then the sorrowing *rākshasas* told their king of Indrajit's death. When the ten-headed Rāvana heard the woeful news, his senses left him, and, as he slowly recovered, he lamented:

'Alas my son, finest of the *rākshasa* warriors and most skilled in the chariot, you that overcame Indra, how could Lakshmana's shaft have laid you low? The three worlds are empty to me now, and when I go to the other world, to the abode of Yama, no Indrajit will perform my funeral rites, for it is I who must perform yours now.'

Then a terrible resentment arose in his heart, and, summoning the still formidable remnants of the demon army, he went forth to the gate where Rāma and Lakshmana stood. When the two great enemies saw each other at less distance than an arrow might fly, Rāvana, bracing himself against his chariot, drew out a fearful dart and hurled it against Lakshmana. Deep the weapon drove into Lakshmana's breast, impelled by the full force of Rāvana's rage. When Rāma saw his brother fallen he exclaimed aloud:

'What care I now for victory when my brother bleeds to death? What fruit in life and what happiness if Lakshmana is gone? Life is no use to me nor Sītā either. Here one may get a wife, there a kinsman, but nowhere can I find my brother again.'

But the wise Sushena broke into Rāma's lament and said:

'Do not grieve thus, for your brother is not dead; Lakshmana still lives.' Then, turning swiftly to Hanuman, Sushena went on, 'Go forthwith to Mount Oshadhi and bring the plant that grows by its southern peak!'

Whereupon once more the hasteful Hanuman sped off and as before returned with the mountain's summit in his hands. Sushena then, uprooting the herb, pressed it and let Lakshmana savour its healing scent. Forthwith Rāma's courageous brother stood up, healed of wounds and pain.

'Despair no more on my account,' he spoke. 'Make the slaying of Rāvana now come to pass.'

Thereupon a shower of arrows flew between the two great adversaries, Rāma and Rāvana, that seemed like a cloud swallowing the rays of the sun. When the gods saw that Rāma stood on the ground, they deliberated and held the fight unfair. From heaven then, they despatched Mātali, the divine charioteer, to Rāma's aid. Then, equally matched and weapon for weapon, the two struck blow and counterblow, until Rāvana, in terrible fury, rained such a storm of arrows on Rāma that the great hero was sore distressed. As the monkey hosts and all the dwellers in the three worlds looked on, they became troubled, for, as the moon is swallowed by Rāhu, demon of the eclipse, so Rāma was overshadowed by Rāvana. Thereupon the great sage Agastya appeared and spoke to Rāma:

'Hear from me how you will conquer your enemy. Repeat constantly the Āditya hymn to the sun, for it destroys all fear, overcomes all evil and brings all that is auspicious. Worship the rising sun, the son of Vivasvan, the creator of light, and adored by all the gods whom he contains.'

With these words, Agastya disappeared as he had come. Then Rāma, looking at the sun, recited the hymn and felt an inner joy. Sipping water thrice, he became purified and, looking forth at Rāvana, began to do battle with him. As chariot wheeled about chariot, the tumult filled the world with fear, but as the sky can be only as the sky, and the ocean only as the ocean, so the battle of Rāma and Rāvana was like none but that between Rāma and Rāvana.

Then Rāma fitted an arrow like a deadly serpent to his bow and in his righteous strength let loose a shaft fit to tear the bowels of the earth. The lord of the *rākshasas*, in all his power and radiance, was torn from his chariot, cast down and slain like the demon Vritra by Indra's thunderbolt.

'Alas my brother,' said Vibhīshana, looking at the demon king, bloody on the ground, 'alas that you did not heed my warning in the folly of your passion; for what I have said would come to pass has now most bitterly come to pass.'

Said Rāma to the sorrowing Vibhīshana, 'A hero either kills his enemies in battle or is killed by them.'

Thereupon Rāma gave orders for Rāvana's funeral rites to be performed, and there came out, from the inner apartments of the palace of Lankā, the demon's heart-stricken wives. The chief queen, Mandodarī, as she gazed upon her husband slain by Rāma, lamented:

'O Rāvana, before whom even Indra trembled, I did not think to see you slain by a mere mortal. When you carried off Sītā, you committed an unworthy deed, for one who does good reaps good, but one who does evil reaps evil. Vibhīshana has gained content while you lie bleeding on the ground. Nor by lineage, nor beauty, nor gentle heart did Janaka's daughter surpass me, nor even equal me. Alas for your folly that you did not see this. She now, united with her husband, will be free from sorrow, while I, my husband slain, wonder that my heart does not shatter to a thousand pieces. O Rāvana, in your great strength you do not need my grieving, but as for me, who am but a woman, my mind revolves in grief.'

When Vibhīshana had performed his brother's funeral rites, Rāma said to the illustrious Lakshmana:

'Instal Vibhīshana as monarch of Lankā, for he is our friend and has helped us.' Then, turning to Hanuman, standing like a mountain at his side, Rāma continued: 'Most noble of the victorious, go forth now into the city and announce to all our victory. Tell Sītā that Sugrīva and Lakshmana and myself are well and that I have slain Rāvana. Do this, Hanuman, give her the joyful news and bring her reply back to me.'

Swiftly the great monkey carried out his task and, after giving Rāma's message to Sītā, he told her: 'Take heart now that all is yours to command. Vibhīshana himself draws nigh, delighted to see you once again.'

Vibhīshana commanded that Sītā, adorned by gentle maidens, be brought in a splendid palanquin to Rāma's presence, and when the great son of Raghu saw her led from the *rākshasa's* palace, the long separation and the sight of her bred joy, sadness and anger, all three, in his heart. As Janaka's daughter drew near to her lord, she hid herself, as it were, in her own limbs and as she stood at Rāma's side, she looked once more at his handsome face with surprise, delight and love.

'My fair one,' said Rāma to her, 'now the enemy is slain and you have been rescued. All that men could do has been done; my wrath is at an end, the insult wiped out, and I am master of myself. How then can I take back one who is open to blame, who has sat on Rāvana's lap and been looked on by his evil eye? You stand before me, like a bright lamp in the face of one come out of long darkness. Go therefore, daughter of Janaka, where you will; I have no desire for you, you may go as you please.'

'Why do you speak such words to me,' said Sītā slowly and in faltering voice, 'words which hurt one's heart, as though you were a common man addressing a common woman? You must believe in me, nor blame me because you suspect all womankind. If the demon has touched me, that was not of my willing; rather the sin lies at the door of fate; but my heart has always remained with you. Why did you not renounce me, when Hanuman first found me? Anger it was that led you on, like an ordinary fellow, and jealousy for my womanhood, but have you not heeded that, when I was young, you, a young man, took my hand in marriage? Are my devotion to you and my character held as naught? Lakshmana, noble Lakshmana, prepare a funeral pyre, for I will enter the flames.'

Lakshmana, who knew his brother's wishes without the need for words, made ready the funeral pyre. Sītā then, after walking slowly round Rāma, paid obeisance to the gods.

'Even as my heart has remained pure,' she prayed, 'and I have never sinned against my lord, so let Agni who sees all and purifies all, protect me on all sides.'

So the daughter of Janaka entered the funeral pyre with heart undaunted. Then, from the midst of the flames, Agni, the fire god, appeared, holding Sītā safely in his arms.

'O Rāma,' said he, 'take to yourself your wife whose heart is pure and free from sin.'

Thereupon the immortals appeared, bearing Dasharatha in an aerial chariot, and Shiva bade Rāma and Lakshmana salute their father. When he had received their homage, King Dasharatha embraced Rāma with great joy.

'Now I understand,' said he, 'how heaven decreed that Rāvana should be slain. Now that the immortals have wrought their plan, long may you rule your kingdom with your brothers.'

'Then let not Kaikeyī incur the wrath of heaven,' answered Rāma, with folded hands, 'nor let harm befall her son Bharata.'

'It shall be so,' said Dasharatha; and then, embracing Lakshmana, he continued, 'You who have stood ever at your brother's side and remained devoted to him only, you, as well as Janaka's daughter, shall gain eternal fame. And you, Sītā, be not angry because you were renounced. Rāma wished for you to be purified and that all good should come of it.'

When he had said these words to all his children, the illustrious monarch returned once more to the world of Brahma, while Shiva, lord of the three worlds, delighted at the outcome, spoke to the devout Rāma.

'Rāma, that you have now seen all the immortals, all of them pleased with your worthy deeds, cannot be in vain. We shall grant you a blessing, so tell us what lies near to your heart.'

'May all the brave monkeys who have fought on my account,' Rāma requested then, 'may they all return from the shades of Yama and receive their lives again.'

'Let it be so,' said Indra, the chief of the gods, and thereupon on every side there arose the trusty apes, anxious every one to salute the son of Raghu.

The gods departed then, commanding Rāma to return to Ayodhyā with his fond and devoted wife; to disband the monkey host and instal himself as king. So, in his haste to see his kinsmen once more, Rāma begged Vibhīshana to prepare the aerial chariot. The kindly Vibhīshana at once made ready the great car of gold, fleet as the wind, and Rāma took his place in it with Sītā on his lap. In the car too were Lakshmana, Vibhīshana and all the mighty monkeys. Driving on as a cloud before the winds, they reached Ayodhyā, and Rāma forthwith dispatched Hanuman to seek out his brother Bharata. Hanuman, assuming human form, went quickly into the city, but ere he reached the palace, he found Bharata in his hermitage, clad in the deerskin, sad and careworn, administering the kingdom on behalf of the golden sandals. Hanuman approached him and greeted him with all respect.

'He whom you mourn,' said Hanuman, 'he whom you think is living in the Dandakā forest with the deerskin and matted locks, inquires this moment for your welfare and soon will appear before you.'

'Truly, if one but lives,' said Bharata, deeply contented at the news, 'joy comes, even if after a hundred years.'

Bharata swiftly spread the tidings, and all the citizens of Ayodhyā, all the womenfolk of the palace, with Kausalyā at their head, came joyously to Bharata's hermitage in Nandigrāma. While the sound of their delight rose up to the very heavens, Rāma approached his brother Bharata, and Bharata, bowing deep before him, himself put the golden sandals on Rāma's feet. Rāma embraced his mother Kausalyā and all his family in the delight of their reunion and then, mounting the royal chariot, he drove into Ayodhyā. As he entered his father's lovely palace the aged Vasishtha came forth and, setting Rāma, with Sītā at his side, upon the jewelled throne, he sprinkled the son of Dasharatha with pure and scented water.

When Rāma was installed as king, the land grew heavy with grain and honey. Calling to him Vibhīshana, Sugrīva, Hanuman, Jāmbavan and all the noble heroes who had been his allies, he, Rāma of radiant fame, honoured them all with jewels and rich gifts. Delighted in their hearts they returned to their own abodes.

So Rāma, with fair Sītā at his side, ruled his kingdom righteously and performed holy sacrifices. The land was free of evil; no calamities befell nor did the old perform the funeral rites of the young.

*He who reads or listens to this tale of Rāma, composed in olden times by Vālmīki, will be freed from sin. To him, prosperity and all good will come.*

## The Price of Greed

ONCE upon a time in the southern forests there dwelt a certain old tiger. Every day he would take a ceremonial bath and, gathering some sacred *kusha* grass in his paw, he would call out to the passers-by as he sat at the edge of a pond: 'Ho there, good travellers, take this golden bracelet!' One day a certain traveller was attracted by greed on hearing the words of the tiger and he thought to himself: This is a lucky chance! But I must not be hasty where a risk is involved for, people say, the result of getting a desirable object from an undesirable source is not good; indeed, even nectar, when tainted with poison, brings about one's death. Still, the search after wealth is always attended by danger, and on this point I have heard it said that no man attains a fortune unless he embarks on an adventure. Then, if he risks everything and survives, he truly gains a fortune. Let me therefore look carefully into this matter. Thereupon he called aloud: 'Where is your bracelet?'

The tiger stretched out a paw and showed it to him, but the traveller said: 'How am I to trust someone with a murderous nature like yours?'

The tiger replied: 'Listen, worthy traveller. Long ago, in the days of my youth, I was most certainly very wicked and I killed

many a cow and many a Brahman. As a result of my sins, my
wife and my children died and now I am without heirs. One
day then a saintly man advised me to practise charity and to
lead a holy life. I followed his advice, so that I am now in the
habit of taking ritual baths and giving presents. I am old now,
and my claws and my teeth have fallen out; how then could you
fail to have confidence in me? Indeed,' the tiger went on, 'I am
so utterly free from all desires that I am willing to give away
this golden bracelet which I hold in my paw to anyone who
wants it. I admit it is difficult to overcome the belief that tigers
eat people, but I, for my part, have studied the laws of religion.
You are a poor fellow, and so I would like you to have this
bracelet. A gift which is given for the sake of giving to one who
can make no return is, they say, the very best of gifts, and especially if made at the proper time and place and to the proper
person. Come and bathe in the pool then and accept the bracelet from me.'

The traveller felt confidence at the tiger's words, but no
sooner did he enter the pool in order to bathe than he found
himself stuck fast in the mud and unable to run away. When the
tiger saw him held deep in the mud he said: 'Oho! you have
fallen into the mud; I will just lift you out of it.'

With these words he gently approached the traveller.

As the traveller was seized by the tiger, he thought to himself:
The fact that he studies the laws of religion is certainly no
reason for having confidence in a villain; indeed, it is the nature
of the person that counts, just as the milk of cows by nature is
sweet. I did not do well in having faith in this murderous beast,
for even the moon is swallowed by Rāhu, the demon of the
eclipse. So fate ordains it, and who can wipe out the decrees of
fate?

With these and other thoughts passing through his mind, the
traveller was killed by the tiger and eaten.

## The Brahman and the Goat

ONCE a certain Brahman began a sacrifice in the forest of Gotama. He needed an animal to offer to the gods, but there was nothing to hand except some dogs which, as well as being animals considered unfit for sacrifice, are always regarded as impure beasts by pious Brahmans. So he went to a village and bought a goat there which was entirely suitable for the sacrifice. As he was journeying back, however, with the goat on his shoulder, three rogues happened to see him.

What a fine trick it would be if we could get this goat from him! the three rascals thought to themselves, so they stood behind three trees, each some distance apart, along the way the Brahman was travelling, and awaited his arrival.

As the Brahman came along, the first rogue stepped out and said: 'Ho there, Brahman! Why are you carrying a dog on your shoulder?'

'This is not a dog,' replied the Brahman. 'It is a goat for sacrifice.'

Farther along the way the next rogue asked him the same question. When he heard him, the Brahman took the goat down from his shoulder, put it on the ground and examined it very carefully. He then put it up on his shoulder again and continued on his way, but he was a little uncertain in his mind. Truly, the minds of even good people waver when they hear the words of the wicked.

Soon he came up to where the third thief stood, and the fellow addressed him: 'Worthy Brahman! Why are you carrying this cur on your shoulder?'

At these words the Brahman thought to himself that it must definitely be a dog and that he had made some mistake. He dropped it straightway, washed himself in a stream to purify himself and went home. The three rogues, for their part, killed and ate the goat.

## The Indigo Jackal

A CERTAIN jackal who lived in a forest fell into an indigo tub as he was wandering about on the outskirts of a town. As he was unable to get out he thereupon pretended to be dead and remained like this till morning. When the owner of the indigo tub saw him, he thought he was dead and so, lifting him out, he carried him some distance away and threw him down. The jackal ran off and entered the forest and then, finding that he was all blue, he began to think to himself: I am now the very best of colours. Why should I not do something now to advance my position in life? After considering the matter, he summoned all the jackals together and addressed them: 'I have been anointed by the holy goddess of the forest; by her own hand I have been anointed with an essence of all the plants in the woods. Look at my colour. From this very day onwards, therefore, everything that goes on in the forest must come under my command.'

The jackals saw that he was indeed a distinguished colour and so, bowing low, with their ears to the ground, they said: 'As your majesty commands, so it will be.'

In this way then the jackal gradually gained sovereignty over all the other animals dwelling in the forest. He surrounded himself with a royal retinue of lions, tigers and the like and, as a result, he began to look upon his own kin, the other jackals, with embarrassment. He therefore contemptuously banished them.

All the jackals were in despair, but one old jackal reassured them. 'Do not be despondent,' said he, 'for, though we have been treated with contempt by this fellow who does not know how to behave, we do know his weak points. I shall arrange something to bring about his downfall. Since these tigers and others are deceived by his mere colour and do not recognize him to be a jackal, they naturally accept him as a king. You must therefore act in such a way that he betrays himself, so follow this plan which I shall describe to you. At evening-time you must all raise a great howl in the vicinity where he is. When he hears that sound he will be bound to give an answering yell on account of his very nature as a jackal. Indeed, people say, one's own true nature is a very difficult thing to shake off; if a dog were made a king, and were hungry, would he not gnaw a shoe?'

The jackals accordingly carried out the plan and surely enough, when the indigo jackal heard their great howl through the twilight, he began to give tongue himself. Immediately then one of the tigers recognized him for what he was and promptly killed him. Such is the fate of a fool who deserts his own side and joins the enemy.

## The Robber Simhavikrama

ONCE upon a time there lived a thief by the name of Simhavikrama who from his very birth had nourished his body by stealing the possessions of everyone about him. In time he grew old and, giving up his thievish pursuits, he thought to himself: What way lies open to me to the next world? To whom shall I go for refuge there? If I seek the protection of the gods Shiva or Vishnu, of what account shall I be to them since they have gods and sages and others as their worshippers? In this case I should pay my devotion to the immortal scribe Chitragupta, for he alone keeps the record of good and bad deeds performed by men and he may save me through his own great powers. He alone executes the business of the Supreme Being, Brahma, and in an instant he writes down or wipes out the entire world which lies in his hand.

With these reflections he embarked upon his devotions to Chitragupta and worshipped him alone, giving alms-food constantly to Brahmans in order to please him.

While the thief was carrying on these practices, Chitragupta came one day to his house in the guise of a guest in order to inquire into his state of mind. Thereupon the thief paid him respect and afforded him alms of food, saying to him: 'Speak these words: "May Chitragupta be pleased with you!"'

Then Chitragupta, still in his disguise as a Brahman, said to him: 'What does Chitragupta avail you if you leave out Shiva and Vishnu and the others? Tell me this!'

When he heard this, the robber Simhavikrama replied: 'What has this got to do with you? I have no need of any deities other than him.'

At this, Chitragupta in the form of a Brahman spoke again: 'If you will give me your wife, then I will say these words.'

Simhavikrama, being delighted at this, answered him: 'Out of love for the deity whom I worship, my wife shall be given to you.'

When Chitragupta heard this he revealed who he was and said: 'I am that Chitragupta and I am pleased with you, so tell me what I may do for you.'

Then Simhavikrama told him in the utmost joy: 'Gracious one, arrange that Death may not touch me!'

'One cannot protect oneself from Death,' Chitragupta said then, 'nevertheless I shall perform a miracle for you, so hear this from me: When Death was burnt up by the lord Shiva who was angry because the great sage Shveta had lost his life, the need for Death became so great on earth that Shiva brought Death to life again. At the command of Shiva though, Death cannot assail any creature who lives in Shveta's hermitage. Now that sage Shveta abides on the other side of the eastern ocean, in a hermitage grove, away beyond the River Tarangini. I shall therefore put you in that place where Death cannot reach you, but you must never return to this nether shore of the Tarangini. Yet if through some carelessness you do come back, and Death should capture you, then I shall help you to escape when you arrive in the other world.' When he had said this, Chitragupta brought the delighted Simhavikrama and placed him in that hermitage of Shveta and then disappeared.

After some time Death approached the nether shore of the Tarangini in order to seize Simhavikrama, but as he stood there he saw no other way to do this except by enticing him with a maiden of heavenly beauty whom he fashioned by his magic power. So that playful maiden went and approached Simhavikrama and enthralled him with her magical gifts and deluded him with the wealth of her own beauty.

After days had gone by she went across the Tarangiṇī with its waves, under the pretence of wishing to see her relations. While Simhavikrama, standing on the bank, gazed after her, she made as if to stumble in the middle of the river, and then she cried as though she were being carried away by the current:

'Look, my husband, I am being swept away! Why do you not rescue me? Are you a jackal by nature or does your name really mean "the strength of a lion"?'

At this, Simhavikrama immediately plunged into the river, but that woman, pretending to be drawn away by the speed of the water, led him in no time to the opposite bank as he followed her in his attempt to rescue her. When he reached there, Death seized him by throwing a noose about his neck; for disaster awaits by the head of those whose minds are consumed by earthly things.

Then, when Chitragupta saw Death leading him, all through his carelessness, into the assembly hall of Yama, ruler of the dead, he spoke to him privately, since Simhavikrama once had pleased him: 'If you are asked whether you will go through hell first or heaven, you should request that you may abide in heaven to begin with. Then, while you are dwelling in heaven you should perform meritorious deeds so that you may gain a permanent place there, and after that you should perform some severe penance to wipe out your sin.' At these words from Chitragupta, Simhavikrama was ashamed and, with head bent down, willingly agreed.

Straightway then, Yama, king of righteousness and of the dead, spoke to Chitragupta: 'Does any portion of credit belong to this thief here or not?'

Chitragupta then replied: 'He is indeed given to hospitality, for out of love for his self-chosen deity he even offered his wife to one who entreated him; therefore he may go to heaven for the duration of a day of the gods, a year of mortal time.'

When Yama heard this he asked Simhavikrama: 'You! Will you enjoy torment first or delight? Tell me!'

Then Simhavikrama pleaded for that delight of heaven in the first place, and thereupon he mounted an aerial chariot which came at Yama's command and, mindful of Chitragupta's words, he travelled up to heaven.

There, because he kept strictly to a vow of prayer and bathed in the holy stream, Mandākinī, and took no delight in the enjoyments of heaven, he gained a further year. In this way by dint of the most severe penances he gradually gained heaven for himself and, after pleasing Shiva, the granter of prosperity, his sin was burnt out and he became pure and holy. As a result of this the messengers sent by Yama to reclaim him were not able to look him in the face; Chitragupta wiped out his debt of sin in the birch-bark book, and Yama remained silent. In this way, Simhavikrama, even though he was a robber, attained heaven on the strength of recognizing good; thus is the perfect nature of wisdom explained.

## A Noble Brahman's Sacrifice

THERE is here on earth a great and beautiful city called Vikramapura where formerly there reigned a king, Vikramatunga by name. There was a sharpness in his sword but not in the rod of justice when he passed his judgements, and he was constantly attached to righteousness but not to women, hunting and other pursuits. And while he was king, only the seeds of evil fell on stony ground, arrows flew from none but the bow of virtue and the only straying was that of sheep from the shepherds of the flocks. There came then one day a handsome hero, Vīravara by name, born in the country of Mālava, to attend upon the king. His wife was called Dharmavatī, his daughter Vīravatī and his son Sattvavara, and this was the extent of his household, while the extent of his retinue was a further trio, the dagger at his hip, the short sword in one hand and a brightly shining shield in the other. Though the extent of his establishment was modest, he used to ask the king every day for five hundred coins for their sustenance. And the king, who had noticed his worth, would give him this pay, thinking to himself: I must find out about his worthiness.

Therefore the king set spies upon him to find out about him and discover what the stout-armed fellow did with so many coins. Every day Vīravara would give a hundred of those coins into his wife's hand for housekeeping; with another hundred he would buy clothes and garlands and so forth; then, after performing his religious duties he would spend a further hundred on offerings to Vishnu, Shiva and other gods while the remaining two hundred he gave to Brahmans, poor people and suchlike. In this way he consumed the whole five hundred every day. He stood at the main gate of the king's palace in the morning and then, when he had performed the rites for the day and other tasks, he came there again in the evening. The spies kept informing the king of this daily routine whereupon he was satisfied and withdrew the spies from their task.

Meanwhile Vīravara, apart from the time spent in prayer and so forth, continued to pass day and night, with his weapons at the ready, at the main gate of the palace. Then there came a mass of clouds, as if trying to overwhelm Vīravara and making a terrible thundering as if unable to countenance his bravery, and there rained a fearful cloudburst like a shower of arrows, but Vīravara, like a doorpost, never moved from the main gate of the palace. King Vikramatunga observed him from the top storey of the palace and, wanting to know about him, he climbed up there again during the night. He called then from the top of the palace: 'Who stands at the main gate of the palace?'

When Vīravara heard him, he answered: 'It is I who stand here.'

At these words the king reflected: Certainly this very noble fellow deserves the highest position, for he does not leave the gate of the palace even when such a torrential cloud pours rain.

Just then the king heard a woman crying pitifully in the distance and he thought: There is no one in trouble in my kingdom, so who is this woman who is weeping?

Thereupon he called out to Vīravara: 'Worthy Vīravara, listen, there is some woman crying in the distance; go find out who she is and what her trouble is.'

When he heard this, Vīravara set out forthwith, his knife at his side, brandishing his sword. As he saw him setting forth, while the lightning flashed and the breaks in the clouds closed with downpours of rain, the king came down from the palace roof and, out of curiosity and compassion, went out after him, unobserved, with his sword in his hand.

As Vīravara kept going on towards the source of the weeping, with the king following secretly behind him, he came outside the town and approached a lake, and there, in the centre of it, he saw a woman crying out:

'Lord, O compassionate one! O hero! How shall I live if I am abandoned by you?'

And when he asked her who she was and what lord she was lamenting, she said to him:

'Know, O Vīravara my son, that I am the Earth. Vikrama-tunga, the king, is at present my rightful lord, but his death will take place for certain three days hence. Where, my son, shall I find again such a lord? This is the reason wherefore I grieve, sorrowing for his very self. For I see the future, what is good and what is evil, with a divine insight when I behold the impending death of the king.'

Then to the goddess Earth who had said these words, Vīravara spoke:

'My mother, if there is any way to save the king, please tell me.'

At Vīravara's words, the goddess Earth replied to him: 'There is an answer to this matter, and it is entirely dependent upon you.'

'Then tell me quickly,' were the Brahman Vīravara's courageous words on hearing this, 'tell me, O goddess, whether my master could be saved through my life or that of my wife and children, so that my birth may be fruitful.'

As he spoke thus, the goddess Earth then answered him: 'There is in this place the goddess Chandikā, whose shrine lies close to the king's palace; if you will offer your son Sattvavara as a sacrifice to her, then the king will live; there is no way out apart from this.'

When the resolute Vīravara heard this speech of the goddess

Earth, he said to her, 'I go, O goddess, and I will perform this straightway.'

'Who else could be so devoted to his master? Go, and may you fare well.' With these words she disappeared; yet the king who had followed had heard all.

Then as King Vikramatunga secretly followed him, Vīravara went swiftly in the night to his own house and there awakened his wife Dharmavatī and told her how, at the behest of the goddess Earth, their child would have to be sacrificed.

'What will serve the master must certainly be done,' said she on hearing the news, 'so wake up your son now and tell him.'

Thereupon Vīravara woke up his child and told him how it had been decreed by the goddess Earth that he should be sacrificed for the king's sake. At these words, the young boy Sattvavara, who had been rightly named as possessing the highest qualities, replied: 'Am I not indeed, my father, destined to gain merit if my life is used in the service of the king? I must assuredly make him a return for his food which I have eaten; so take me therefore and offer me as a sacrifice to the goddess on his behalf.'

When Vīravara heard his child saying this, he answered him without distress: 'Truly indeed you are my offspring.'

When King Vikramatunga who was standing outside heard it, he reflected: They, all of them, are equally brave.

Then Vīravara took his son Sattvavara on his shoulder and his wife Dharmavatī placed her daughter Vīravatī on her back, and the two of them went in the night to the shrine of Chandikā; and again King Vikramatunga secretly followed behind them. Sattvavara was then set down from his father's shoulder at that place and although he was yet a child he was so full of courage that he bowed down before the goddess and addressed her:

'O goddess, may our master live through this sacrifice of my head, and may King Vikramatunga rule this earth without difficulties besetting him.'

When his son had said this, Vīravara exclaimed: 'Well done, my son!' and, drawing his sword, he cut off his head and offered it to the goddess Chandikā, saying, 'May the king be fortunate.' For indeed, those who are devoted to the welfare of their masters take no delight in their own lives or those of their sons.

Then a voice from heaven was heard to say: 'Well done, Vīravara, you have bestowed life on your master through the vital breath of your son.'

As the astonished king saw and heard all that went on, Vīravara's little girl, Vīravatī, ran up to her brother, embraced his head and kissed it, weeping: 'Alas! my brother!' so that her heart burst and she died. When she saw that her daughter was also dead, his wife Dharmavatī then spoke to Vīravara, folding her hands in grief:

'We have carried out what was necessary to save the king; so grant me now permission that I may ascend the funeral pyre with my two children whom death has taken. As my little girl in her innocence has died out of grief for her brother, what pleasure is there in life for me, now that my two children are gone?'

As her words were determined, Vīravara then said to her: 'Do as you will! What am I to say? For there is no joy for you, my faultless wife, in this life, which would be nothing but grief for your children; therefore wait until I make ready the funeral pyre.'

With these words, he made up a funeral pyre with the wood which formed the enclosure of the shrine and put the bodies of the two children on it and set it alight. Then his wife Dharmavatī bowed down at his feet, saying: 'May you, my worthy husband, also be my husband in the next world, and may the king enjoy prosperity.'

With that the worthy woman fell down in the funeral fire with its crest of flames as easily as if it had been a cool lake. When King Vikramatunga saw this as he stood hidden, he remained absorbed in thinking, How shall I ever repay my debt to these people?

Then Vīravara, he of firm resolve, reflected to himself: I

have carried out my duty towards my master, for the voice from
heaven was clearly heard by me. My master has been requited
for what I have eaten and enjoyed, and now I have spent everything in support of my beloved family. Life would not be pleasant to me supporting myself alone, so why should I not offer
myself also as a sacrifice to the mother goddess?

So Vīravara, steadfast in character, made his resolve and
worshipped Chandikā, the goddess who granted wishes, with
a hymn of praise: 'O great goddess, I bow before you! You who
give refuge to them that worship you, raise me who am fallen
into the mire of this existence and who have come to you for
protection. You are the giver of life to living creatures, and in
the beginning of creation you were beheld by Shiva himself,
making radiant the whole universe as you shone out with
splendour too bright for the eye. You were praised by that very
trident-bearing god Shiva as well as by the other gods, and as
they praised you, O gracious one, immortals, seers and men
received their requests over and beyond desire, and receive
them yet. Therefore be kindly to me, O giver of blessings. Take
this sacrificial offering of my body; may my master the king
enjoy prosperity.'

Thus he spoke and was about to cut off his own head when
from the face of the skies there came a bodiless voice:

'Do not act so hastily, my son, for I am well pleased with this
integrity of yours; therefore ask from me whatever blessing you
desire!'

When he heard this, Vīravara said: 'If you are pleased, O
goddess, then may King Vikramatunga live another hundred
years and may my wife and children also live.'

When he had made this wish, the voice from heaven was
heard once more: 'So let it be!'

At that very moment Dharmavatī, Sattvavara and Vīravatī
rose up alive with their bodies unharmed. Thereupon the delighted Vīravara led them who had been brought to life by the
favour of the goddess back to his own home and then he returned to the king's gate.

And when King Vikramatunga had seen all this, in astonishment and amazement he returned and mounted to the top of his palace unobserved. 'Who is at the palace gate?' he called down. When Vīravara who was standing underneath heard him, he answered him, 'It is I who stand here. I went along to look for that woman, but like some goddess she eluded me and disappeared as soon as I saw her.'

When King Vikramatunga heard this, after seeing the whole adventure in all its full miraculousness, he thought to himself, alone in the night: Without compare assuredly is the magnificence of this fellow who performs such praiseworthy things and yet gives no account of them. The ocean which is deep and broad and full of life does not match this man who is unmoved even at the blast of a mighty wind. How can I make any return to him who secretly gave me my life this night by the sacrifice of his wife and children?

As the king pondered over this and other things he descended from the roof of the palace and going inside he passed the night in contentment. The next morning then, as Vīravara stood in the great assembly, he related the wonders that had taken place the night before. Then, while everyone was praising Vīravara, the king invested him and his son with the turban of honour. And he gave him many districts, horses, jewels, elephants and ten millions of gold coins as well as a salary sixty times as great as before. That instant the Brahman Vīravara became like a king with the lofty umbrella of royalty and, together with his family, achieved contentment.

## Kitava the Gambler

THERE was formerly in this city of Ujjayinī a roguish
gambler named Kitava who was known as the terror of
the gambling dens. Since he perpetually lost while the
other gamblers used to win in play, they would give him every
day a hundred cowrie shells. With these he used to buy wheat
flour from the market-place and at the end of the day he would
make little cakes by mixing it up in a skull with water and, going to the cemetery, he would cook them on a funeral pyre and
then return and eat them, smeared with *ghee* from the sacred
lamps of the shrine of Shiva. And there, in the courtyard of the
temple of Shiva, he would always sleep at night, on the ground,
with his arm as his only pillow.

One night there in that temple of Shiva, while he was looking
at the images of the group of mother goddesses and lesser deities,
the thought flashed up in his mind from his nearness to the
spells:

Why do I not devise some cunning plan to gain riches?
If it succeeds, well and good; but if it fails, what harm will I
suffer?

With these thoughts he invited the mother goddesses to
gamble, saying: 'Come along, I will play with you here. I, a

keeper of the gambling table, will make the throws and whatever I win must be handed over.' As they remained silent when
he said this, Kitava then made a throw after staking some bright
cowries. For when the dice are being thrown, gamblers everywhere accept the rule that whoever wishes to stay out of the
game must say: 'I refuse!', and in gambling this is a condition
agreed to by gamblers everywhere.

When he then had won a quantity of gold, he said to those
goddesses: 'Give me the money I have won as agreed by you.'
As the deities did not say anything, although addressed by the
gambler many times, he then spoke to them angrily:

'If you intend to remain silent, I shall treat you as one usually
treats a gambler who stonily refuses to pay what he has lost; I
shall saw through your limbs with a saw sharp as the teeth of
Yama, for I have no regard for anything.'

And as he said this he took up a saw and ran upon them,
whereupon the deities gave him the gold which he had won.
Then, after losing it the next day at play, he came back again
at night and once more forcibly extracted money from the host
of mother goddesses by dicing.

So he kept on doing every day, until the goddess Chāmundā,
uneasy in her mind, spoke to those mother deities:

'When a person is invited to play, if he says: "I am keeping
out of the game", he cannot be drawn in; such is the custom in
gambling, O mother deities. So by saying this to anyone who
invites you to play, you will cast him out.'

When the goddesses were told this by Chāmundā, they took
it to heart and when the gambler came that night inviting them
to play at dice, all the deities said to him: 'We are staying out of
the game.'

Kitava, now that he had been rejected by them, called upon
their lord Shiva himself to play. But the god, realizing that he
was taking an opportunity of forcing him to play, said to him:
'I stay out of the game.' Indeed, even the gods act like feeble
people in front of a dangerous rogue who thinks only of himself
and boasts that nothing can harm him.

Thereupon Kitava became depressed at his gambler's scheme being foiled by a knowledge of the rules of play and he thought to himself: Alas! I have been frustrated by the gods who have learnt the conditions of gaming, so now I will seek refuge with the lord of gods himself. Reckoning thus in his heart, Kitava grasped the feet of Shiva and declared to him:

'I exalt you, seated with your limbs naked, your cheek resting on your knee; you who once lost your crescent moon, your bull and your elephant skin at gaming to your wife Devī. You, at whose mere wish the gods gave their great powers, who, without wants, possess only matted locks, ashes and the skull, how can you now be mean towards me who have few merits, and, for no great reason, aim at deceiving me thus? You have three eyes, like the dice, so I am like you; on your body there are ashes, as also on mine; and just as you eat from a skull, so do I; have mercy on me! Now that I have talked with you, how shall I talk with gamblers again? Deliver me who am in distress.'

With this and more the gambler praised Shiva until the god, being pleased, said to him:

'Kitava, I am pleased with you, do not be disturbed! I will provide you with material comfort; just remain here with me!' So at the command of the god, the gambler remained there then, enjoying an abundance of comforts produced by the favour of the god.

Then one time there came some *apsarases*, heavenly maidens, to bathe in the holy waters of Shiva at night, and when the god saw them he ordered him:

'Quickly take the clothes which all these heavenly maidens who have come to bathe have thrown on the bank, and bring them here. Then, until they yield you the young *apsaras*, Kalāvatī, do not let them have their clothes back.'

With this order from Shiva, Kitava went and took the clothes of those maidens bathing, with their glances like heavenly deer.

'Let go! Leave our clothes alone! Do not leave us naked!'

But as they cried, he answered them with the power of Shiva:

'If you will give me the maiden Kalāvatī, then I will let you have your clothes back, not otherwise!'

When they heard this and saw that he was difficult to approach, they remembered a curse of Indra on Kalāvatī and so they assented. And then they gave Kalāvatī to Kitava with due rite, and he gave up their garments.

When the *apsarases* had gone away, Kitava remained with his Kalāvatī in a house fashioned at the command of the god. And Kalāvatī went by day to heaven to attend upon the king of the gods, but at night she always returned to her husband.

Once as she was affectionately saying to her husband: 'On account of my obtaining you, my beloved, Indra's curse upon me has been as a blessing,' she was asked by Kitava about the cause of the curse, whereupon the divine bride Kalāvatī related:

'When I saw the gods in the garden, I praised the pleasures of mortals and I made little of the pleasures of divine beings as giving delight to the eye only. When the king of the gods heard this he cursed me saying: "Go, you shall enjoy those earthly delights by marrying a mortal." For this reason the marriage which we both agreed to came about. Tomorrow then I return to heaven after a long time; but you are not to be despondent. For the heavenly nymph Rambhā will dance a new piece in the presence of the god Vishnu, and I, my love, must remain there until it finishes.'

'I will see this dance in secret, so take me there!' said Kitava, who had been spoilt by her love, but when Kalāvatī heard this she said to her husband: 'How would this be proper for me? For the emperor of the gods would be angry if he knew about it.' In spite of these words he made a compact with her, and then Kalāvatī out of love agreed to take him there.

The next morning, after hiding him by her magic power in the lotus flower in her ear, she brought Kitava to the palace of the lord of gods. When Kitava saw it, with its gates adorned with celestial elephants, and fair with the pleasure-grove built by Nandana, he was delighted and thought himself a god. And he saw there, in the court of Indra, the conqueror of Vritra, surrounded by the gods, the wonderful festival of Rambhā's dance, accompanied by the songs of the heavenly maidens. And he heard all the instruments played by Nārada, the divine musician, and others; for what cannot be gained in this world if the supreme god Shiva is favourable to one?

Then, at the end of the spectacle, an actor came forward and, dressed up as a divine goat, began to dance heavenly dances. When Kitava saw it he recognized it and thought to himself: Indeed, I have seen a goat such as this in Ujjayinī, a mere animal, and here it is as a dancing mime in the presence of Indra. This must surely be some strange and wonderful trick of the imagination which the gods have worked. While Kitava turned these thoughts over in his mind, Indra left the place at the end of the goat-mime dance. Then Kalāvatī in her delight took Kitava, still lurking in her ear-lotus, back to his own abode as well.

And on the next day when Kitava in Ujjayinī saw, as he thought, the divine actor in the form of a goat, he addressed him arrogantly:

'You there! Dance before me as you danced for Indra! Otherwise I shall deal roughly with you; so show your dance, O mime!' When the astonished goat heard this it remained silent, thinking: How could this man have imagined this? Then as the goat, even though obstinately addressed, still did not dance, Kitava beat it about the head with clubs.

Then the goat went to Indra, with its head bleeding profusely, and told him what had happened. Indra then, through his profound insight, learned that Kalāvatī had brought Kitava to heaven for the dance of Rambhā and how the dance of the goat-mime had been seen by that sinful fellow; and he summoned Kalāvatī and cursed her:

'Since, in order to make him dance, the goat has been reduced to such a state by this mortal whom you out of passion secretly brought here, therefore go! You shall become a wooden

image on the pillar made in the temple in the city of Nāgapura
by the King Narasimha.'

When he had said this, Indra was humbly entreated by the
mother of Kalāvatī, whereupon out of pity, he settled a limit to
the curse:

'When the temple, completed after many years, shall be destroyed and levelled with the earth, then shall the curse be
atoned.'

At this curse of Indra and the termination set to it, Kalāvatī
came tearfully to Kitava and told him, blaming him. And when
she had given him her ornaments she disappeared and going to
Nāgapura, she entered the statue of the woman on the top of the
pillar in the temple.

Kitava, for his part, was struck by the poison of separation
from her, neither seeing her nor hearing her, and he rolled
about, stupefied, on the ground.

'Alas! What I knew was a secret and I have revealed it in my
foolishness; for how could there be self-restraint in people of my
kind who are unsteady by nature? And now this cruel separation has befallen me.'

This and more the gambler lamented as he regained his
senses. Then he suddenly had an idea: This is not the time of
despair for me; if I take a firm hold of myself, why should I not
strive to put an end to her curse?

With these reflections the rogue deliberated and, donning the
garb of a wandering ascetic, he went to Nāgapura, wearing his
hair matted, the antelope skin and the string of beads. There in
the forest, at four cardinal points outside the town, he deliberately hid four pots containing his beloved's ornaments, covering
them up in the ground. Then, within the city at night, in the
courtyard in front of the god, he purposefully dug a pot into the
earth, full of five precious gems, gold, diamond, sapphire, ruby
and pearl.

When he had done this he remained there on the bank of the
river in a hut which he had made, performing there a gambler's
penance, engaged in false meditation and the saying of prayers.

Every day he would bathe three times, eating alms-food which he washed with water on a stone, and he gained a reputation as a great hermit.

Eventually the king got to hear of him, but though invited by him he did not go to his presence and so the king came to see him. The king remained there a long time in conversation, and when he was about to go that evening, a female jackal suddenly uttered a howl from afar off. When he heard it the gambler, in his guise of ascetic, smiled, but when asked why he did so, he answered: 'Never mind!'

But as the king asked him obstinately, the wily fellow answered him thus:

'"In the forest to the east of this city, in a grove of reeds, there is a pot full of jewel ornaments; therefore you should take it." This was what the jackal told me, your majesty, for I understand the noises of animals.' After saying this, he conducted the king, assailed by curiosity, to the place, dug up the ground and drawing the pot out, he gave it to him. Then, when the king had received the ornaments, he gained confidence in the selfless ascetic whom he considered knowledgeable and truthful. So he conducted him to his hermitage and after bowing at his feet, he returned by night with his ministers to his palace, praising his good qualities.

In this way, as time went on, the rogue, by pretending to know the cries of animals, handed over to the king who kept coming, the other three pots of jewels from the other quarters. Then the king, with the townspeople, the ministers and the king's wives, paid his devotions to that ascetic alone and became, as it were, entirely ensnared by him.

One day when that wretched ascetic was brought by the king for a moment to the temple, he heard the caw of a crow in the market-place and he said to the king:

'Did you hear the sound of the crow? "There is in this very market-place in front of the god a pot of fine gems buried in the ground; so why do you not take it?" This is what the crow said, so come along and get it for yourself.'

With these words that false ascetic brought the king along and presented to him the pot of fine jewels which he took out of the ground. Then, in his great contentment, the king himself took that fraudulent expert by the hand and entered the temple.

There the ascetic knocked against the wooden image on the pillar which was inhabited by his dear Kalāvatī and saw her. And Kalāvatī, remaining in the form of that wooden image, was grieved when she saw her husband there and she began to weep. When the king and his attendants in astonishment and dismay saw this, they asked the false seer: 'What is this, gracious one?'

Then the rogue, as if depressed and perplexed, said to them: 'Come to your palace; there I will tell you what cannot be told here.'

When he said this he accompanied the king to the king's palace and told him:

'Since this temple was built by you in an inauspicious place and day, bad fortune will be yours, three days hence. At the sight of you therefore, the maiden on the pillar burst into weeping; if you are concerned then about your body, look into this, your majesty, and have this temple speedily razed to the ground this very day. Then make an abode for the gods somewhere else at an auspicious time and place; let the evil omen be allayed and let there be prosperity for you and your kingdom!'

When the king was thus addressed by him, in his fear he issued commands to his subjects and that very same day he razed the temple to the ground. Then he began to build another temple in another spot; for indeed, when rogues gain the confidence of their masters, they deceive them with their tricks.

When the roguish Kitava had thus accomplished his plan, he put aside his ascetic's garb and fleeing away, journeyed to Ujjayinī. And Kalāvatī, freed from her curse now that the temple had been levelled to the ground, met him on the way and in her delight she comforted him and went to heaven to see Indra. Indra, however, was astonished when he learnt from her of the guile of her husband the gambler, and he laughed and was pleased. And Indra sent Kalāvatī to bring the gambler to heaven. There, the emperor of the gods, being pleased and thinking highly of his cleverness and perseverance, gave him Kalāvatī and made him one of his own attendants. So Kitava, heroic as a god, lived in happiness in heaven with Kalāvatī, through the grace of Shiva.

## The Fate of the Vulture

THERE is a mighty silk-cotton tree on a hill named Vulture
Peak near the banks of the River Bhagīrathī. In a hollow
in that tree an old vulture named Jaradgava had made
his home. As a result of a misfortune he had lost his eyes, so out
of pity the other birds who lived in the tree used to give him
bits of their own food for him to live on.

Then one day a cat named Dīrghakarna, or Long Ears, approached the tree in order to eat the young birds. When they
saw him, the nestlings were terrified and made a great noise.
Jaradgava heard it and he called out: 'Who goes there?'

When Long Ears saw the vulture he said to himself: Alas! I
am as good as dead; I am so close to him now that escape is impossible. Things must take their course however, so I shall just
approach him.

With these thoughts in mind he went up to the vulture and
said: 'Noble sir, I greet you.'

'Who are you?' the vulture asked.

'I am a cat,' was the reply.

'Then get yourself far away,' the vulture said, 'for if you do
not, I shall kill you.'

'Please listen to what I have to say,' went on the cat, 'and then, if I am to die, let me die. Is a person to be killed or to be respected just on account of his birth? Surely one looks into his actions first before deciding to kill him or to respect him.'

'Tell me then,' replied the vulture, 'what exactly is your business?'

'I live here on the banks of the Bhagīrathī,' said the cat, 'and every day I take a ritual bath; I have renounced eating meat, I lead a celibate life and I practise fasts at the new and full moon. The birds have always praised you in front of me because you know the laws of religion, and I have confidence in what they say. For this reason I have come here, so that I may learn about *dharma*, the law of righteousness, from you who are old in years and in knowledge. Yet, it seems, such is your knowledge that you were ready to slay me, a guest. Surely the duties of a householder demand that even if an enemy comes to one's home, fitting hospitality should be given to him; indeed, the tree does not withdraw its shade from the woodcutter come to hew it down. If there be no food, at least a guest should be honoured with kind words, for, it is said, a seat of grass, a shady corner, cool water and, as a fourth gift, kindly words, these things are never missing in the house of a worthy person. Good people show compassion even to the worthless, just as the moon does not withdraw her beams from the dwelling of an outcast. If a guest turns disappointed from a door, he transfers his sins to that house, and whatever merit there be in the house, that he takes with him on his way. Even the lowliest who come to the house of the highest caste must be honoured in true wise, for a guest is like a messenger of the gods.'

'Cats are fond of meat,' said the vulture, 'and here there are young birds dwelling; that is why I have spoken to you thus.'

When the cat heard this he touched the ground and he touched his ears in deep respect and said: 'I am free from such desires and since I have learnt the laws of righteousness I have undertaken this arduous fasting. Even though the religious books disagree on many points they all say the same about this: to do no injury is the highest of duties. For when one eats the flesh of another, what a difference there is between the two! The one gains but a moment's contentment, while the other loses his life. The torment that rises in a man at the thought that he is going to die, that torment should make others, if they can feel it, seek to save his life. Who would commit such a terrible crime for the sake of his belly, when it can be filled by the fruits that grow in the forest?'

In this way the cat gained the vulture's confidence and he remained there, dwelling in the hollow of the tree. Then, as the days went by, he would climb up and catch the young birds. Bringing them back to the hollow, he would feast on them every day. When the birds discovered that their young had gone they were sorrow-stricken and began to search around. As soon as he saw this, the cat crept quickly out of the hollow and fled. Thereupon the birds, in the course of their searching, discovered the bones of their young ones in the hollow of the tree.

'Jaradgava himself must have eaten them!' Such was the conclusion they came to and they killed him. True indeed is the well-known saying: shelter should not be given to one whose family and whose habits are not known; for through the crime of the cat, Jaradgava, the vulture, lost his life.

## Curd Ears

ON a mountain known as Million Peaks in the northern country there lived a lion, Great Strength by name. As he slept in his cave in the mountain, a certain mouse was in the habit of gnawing the tips of his mane. When he saw that the tips of his mane were being chewed up, he became angry and, as he was unable to catch the mouse which fled into its hole, he began to reflect: An enemy which is insignificant cannot be brought low by strength; a warrior of his own size must be appointed to kill him.

With these thoughts in mind he went to a village and by enticing a cat named Curd Ears with bits of meat and so forth, he brought him carefully back to his den and kept him there. From this time on, the mouse did not come out of his hole through fear of the cat, and so the lion slept on comfortably with his mane undamaged. Whenever he heard the mouse making a noise, he was more than ever careful to feed the cat with bits of meat.

Then one day the mouse could stand his hunger no longer and, as he sallied forth from his hole, he was caught by the cat, killed and eaten. After this the lion no longer heard the sounds made by the mouse and so, as the cat had lost his usefulness, he became careless about giving him food. Indeed, people say, a master should never be made feel independent by his servants, for when a servant makes a master independent, he may fare badly, just as the cat, Curd Ears, did.

## The Hare in the Moon

ONCE upon a time, even though it was the rainy season, no rain fell. A herd of elephants, being tormented by thirst, said to their leader: 'Sir, there seems to be no way for us to save our lives. Here there is but a tiny pond fit for small creatures only; we are all but blind for want of a bathe. Where shall we go? What shall we do?'

Thereupon the leader of the herd went a short distance away and showed them a lake full of clear water. As the days went by, however, the hares that dwelt on the banks of the lake were crushed under the trampling feet of the elephants. A hare, Shilīmukha by name, therefore summoned all the hares together and uttered the following thoughts: 'This herd of elephant is troubled by thirst and will certainly come here every day; as a result, our tribe will be destroyed.'

At this an old hare named Vijaya spoke up: 'Do not despair! I shall put a stop to this.' With this promise he set off and as he went along, he thought to himself: How shall I address the herd of elephant when I approach them? For, as people say, an elephant can kill with a mere touch; a snake can kill just by smelling one; a king has only to smile for a man to die, while a rogue can slay even when paying one respect. I shall therefore climb to the top of this mound and address the leader of the herd.

He did so, and the leader of the herd said to him: 'Who are you? From where have you come?'

'I am an ambassador,' the hare replied, 'sent by the worshipful Moon.'

'Then state your business,' said the leader of the herd.

Vijaya then went on: 'Listen, most mighty elephant; even though weapons be raised against him, a messenger never speaks falsely; indeed, because his life is held sacred, he always relates the truth. I therefore speak at the command of the Moon. Listen! This is what he says: "You have acted wrongly in scattering the hares, the guardians of the Moonlake. These guardians, these hares, they are my subjects, and for this reason I am known among men as the Hare-in-the-Moon." ' [1]

At these words of the hare, the leader of the herd was terrified and said: 'We did this out of ignorance. I will not go there again.'

'In that case then,' replied Vijaya, 'make your bow to the worshipful Moon who is trembling with anger in the lake here; ask his pardon and go!' So he took the leader of the herd by night and showed him the rippling reflection of the moon in the waters and made him bow the deepest of bows.

'Great lord Moon!' said the hare, 'this fellow did what he did through ignorance; let him be forgiven!' With that he sent the leader of the herd about his business. Even the powerful, people say, can be overcome by pretending there is a higher authority; in this way the hares lived happily through pretending to be subjects of the Moon.

[1] The Hindus see the shape of a hare in the moon; not as with us, a face.

## A Tale of Two Rogues

THERE was a city as magnificent as one could desire named Ratnapura, and in it two rogues named Shiva and Mādhava dwelt. These two, with a following of several other scamps, had for a long time been robbing all the rich folk of the town by means of their wily practices. There came a time once when the two held council among themselves as follows:

'The whole town has been thoroughly plundered by us. Let us go then without further ado and live in the city of Ujjayinī, for there, they say, the king's household priest, Shankarasvāmin, is very rich, and by using the money we shall take from him we can discover the delights of the affections of the Mālava girls. The Brahmans there say he robs them of half their fees with a frowning face, for though his riches would fill seven jars, he is a niggard. The story goes too that the Brahman has a jewel of a daughter as well, so she can be got out of him also in this affair.'

So they deliberated, and when they had decided on an agreed plan, those two rogues, Shiva and Mādhava, thereupon set forth from the city. When they had discreetly gained Ujjayinī, Mādhava together with his retinue, dressed in the guise of

Rajputs, remained in a village somewhat outside the town.
Shiva, on the other hand, who was skilled in every form of
deceit, disguised himself as a religious ascetic and entered that
city ahead by himself. There he made his abode in a hut on the
bank of the River Siprā, laying out lumps of clay, *kusha* grass,
an alms-bowl and a deerskin, so that they should be seen, and
every morning he used to smear his body thickly with clay,
practising, as it were, to a fine measurement the smearing with
filth that was to be his lot in the lowest hell. He used also to
remain for a long time head downwards, plunged in the waters
of the river, as if trying out the descent about to befall him as a
result of his evil deeds. When he came out from bathing then
he would stand for a long time with his face held upwards towards the sun, showing, as it were, that he deserved to be im-
paled on the stake. Then, as he went to the temple, holding a
bunch of *kusha* grass and muttering his prayers, he would sit
there in an ascetic pose with a sly and cunning face. As by his
trickery he gained a place in the hearts of worthy people, so,
taking fresh flowers, he made his devotions to Vishnu, and once
he had made his offerings, he again became intent on muttering
his false prayers and so prolonged his devotions as though he
were fixing his attention on the evil ways of the world. In the
latter part of the day then, dressed in the black antelope skin,
he would seek alms, wandering about the town as if deceiving
it with his wily glances; then silently taking three handfuls of
food from the houses of Brahmans, and bearing his staff and
antelope skin, he would divide the food into parts; part he
would give to the crows, part to the person he had visited and
with part he filled his leathery stomach. For a long time he
would hypocritically turn over his beads, muttering away, as
if he were quietly counting up his own sins, while at night he
would stay alone inside his hut, reflecting upon the attitudes of
the world, even to the subtlest points. As he thus performed
every day a harsh, though deceitful, penance, he gained favour
on all sides in the minds of the citizens. The rumour arose
everywhere:

'Truly here is an ascetic who has overcome attachment to worldly things,' and all the people were humble in their devotion to him.

In the meantime the other rogue, his friend Mādhava, hearing about him through his spies, also entered the town and when he had made his abode in a distant temple, still in the disguise of a Rajput, he went down to the bank of the Siprā to bathe. After he with his followers had bathed, he noticed Shiva in front of the temple, intent on his prayers, and he fell down at his feet, making the very deepest of bows. He said out in front of the people:

'There is not another ascetic such as this, for I have seen him many times wandering about to the holy places.'

Shiva, however, though observing him, slyly kept his neck stiff and remained just as he was, while Mādhava went back to his lodging. Then the two of them met at night and, after eating and drinking together, they deliberated as to how the remainder of the plan should be carried out. Shiva then went back leisurely in the last watch of the night to his own hut, while Mādhava, when morning came, gave the following order to one of his gang:

'Take this pair of garments and go to Shankarasvāmin, who is the king's household priest, and give the priest this message, showing him the greatest respect: "There is a Rajput, Mādhava by name, who has come here from the Deccan through being driven out by his kinsmen, and he brings with him a very rich inheritance. He has been accompanied here by some few other Rajputs and he wishes to enter the service of your king here; for this reason I have been sent to see you, O treasure-house of glory."'

When the rogue employed by Mādhava had been thus commanded he went off, carrying the present to the house of the priest, and when he arrived there he gave the present at a discreet moment and delivered to him Mādhava's message in fitting manner. The priest, for his part, believed in it on account of his greed for the presents and with future favours in mind;

for a bribe is the best medicine for attracting those in search of gain. When the scamp had returned, Mādhava himself took the opportunity on the next day and went to see the priest, surrounded by his aides, a wandering band of rogues who pretended to be Rajputs, distinguishing themselves by carrying wooden clubs. Being announced in advance he went in to greet the priest, and his visit was welcomed in turn with charming cordiality. Thereupon Mādhava stayed with him a while, engaged in conversation, and after being given permission to leave he went back to his own abode.

The next day he again sent as a present a pair of garments and once more he visited the priest and said to him:

'For the good of my retainers I should like to enter service, which is the reason I have come to you; incidentally, we also have money.'

When the priest heard that from him he thought about what he might gain and straightway promised to Mādhava what he had asked for. Thereupon he went to the king and informed him about the matter, and the king for his part agreed to the whole affair out of respect for the priest. The next day then the priest brought Mādhava together with his followers and presented them to the king with all due respect. The king, moreover, when he saw Mādhava in all appearance like a Rajput, welcomed him with honour and assigned him a post. So Mādhava remained there serving the king, and every night he met with Shiva for a deliberation. The priest, moreover, with the greed of one who hopes to live on gifts, invited Mādhava to live in his own house, so the rogue repaired there to his house together with his attendants; just as the rat that made its abode in the trunk of the tree was the cause of its destruction. Then, when Mādhava had made a decoy out of ornaments fashioned from fake jewellery, he placed them in a casket in the priest's strongroom, and every now and then he would open the casket and by a pretence of half letting the priest see them, he snared his mind with those ornaments just as the mind of a cow is snared with grass. When the priest's confidence had been completely

gained, he then pretended to become weak in health by losing
weight through taking less and less food.

After a few days had passed, that king of rogues said, in a
weak voice, to the priest who was standing near his bedside:

'Since some evil condition now exists in this body of mine,
please, O most noble of Brahmans, bring some worthy Brahman here upon whom I may bestow my riches for the sake of
my welfare both here and in the world hereafter, for what value
have riches in this unstable existence for a thoughtful man?'

When the priest, that slave to favours, had been thus addressed by him, he promised that he would do so, whereupon
the other fell at his feet. Yet no matter what Brahman that
priest brought, Mādhava would not confide his trust in him,
but insisted that he wanted someone more distinguished. One
of the rogues who stood in attendance, when he observed this,
spoke as follows:

'Seemingly an ordinary Brahman does not particularly
please him, but we should find out whether the great ascetic
Shiva who is now living on the banks of the Siprā would please
him or not.'

When Mādhava heard this he said with an assumed distress
to the priest: 'Please be gracious and bring him, for there is not
another Brahman like him.'

At these words of his the priest then went to where Shiva was
and found him, motionless, pretending to meditate. He went up
to him and walked round him to the right and thereupon the
rogue for his part slowly began to open his eyes a little, and the
priest, after making a most humble bow to him said:

'If you would not be angry, there is something, O Master,
that I would like to tell you. There is a very rich Rajput here
who has come from the Deccan; Mādhava is his name and he,
being in dire health, is anxious to give his riches away, so, if you
are agreeable, he will give you that treasure which flashes with
ornaments made of all kinds of priceless jewels.'

When he heard this, Shiva gradually abandoned his silence
and said:

'O Brahman, what use are riches to a pious ascetic like myself who just lives on alms?'

Then that priest once more spoke to him:

'Do not speak thus, O great Brahman, for do you not know the stages in the life of a religious man? By taking a wife, performing the rites to the gods and ancestors in his house, as well as being hospitable to guests, he is using his riches to carry out his threefold aims in life, for the householder's is the best among all the stages.'

Then Shiva replied:

'How would I be able to find a wife, for I could not marry someone from just any family.'

When the greedy priest heard this and thought about the money he would be able to enjoy at his leisure, he seized his opportunity and answered him:

'There is in this case a girl, my daughter, Vinayasvāminī by name; she is very beautiful and I would give her to you for a wife, and I will look after whatever riches you receive as a present from Mādhava, so why not enter into the stage of householder?'

When he heard this and had thus got the object of his desire, Shiva said:

'O Brahman, if you insist on this, then I will do as you say; yet I am an ascetic who am but a fool in the matter of gold and jewels. I rely therefore upon your word, so do then as you think best.'

When the priest heard what Shiva had to say, the fool readily agreed and straightaway brought him to his own house. There he explained to Mādhava what had been done and was praised by him. Accordingly then he gave to Shiva his own daughter who had been carefully brought up, just as if he were squandering his riches in his stupidity.

Once then that he was married, he brought him on the third day after the ceremony to Mādhava who was pretending to feel low, whereupon Mādhava said to him, quite truthfully: 'I greet you whose asceticism is beyond understanding,' and getting up thereupon fell at his feet. As had been arranged, he gave him all that was brought from the treasure-room, namely the casket with its abundance of fake jewellery. When Shiva received it, he entrusted it to the hands of the priest, saying:

'I know nothing about this, it is for you to find out.'

'I gave you a promise before about this so have not a care about it,' was the priest's answer as he immediately took hold of the casket. When Shiva then had departed together with his wife, after conferring his blessing, the priest took the casket to his own strong-room and stored it there.

The following day, however, Mādhava, gradually leaving aside his pretence of sickness, claimed that he was getting better as a result of the generous gift he had made, and he praised the priest who came to his side, saying:

'Through you and your piety I have recovered from this great misfortune.'

He further openly struck up friendship with Shiva, declaring that his body had become whole through his great powers. Shiva, moreover, as the days went by, said to the priest:

'For how long am I to enjoy the comforts of your house in this way? Why do you not buy those ornaments for some money and if they are worth much, give me a fitting price?'

When the priest heard this he thought that they were beyond price, and agreeing, he gave to Shiva his entire riches to pay for them. On that account he made Shiva sign a receipt with his own hand and he himself did the same, considering the treasure to be worth more than his own money. The priest then, taking the countersigned receipt in his hand, went his own way, while Shiva, for his part, set up a separate household. So the two of them, Shiva as well as Mādhava, remained there, enjoying to their content the priest's riches.

As time went by the priest, in need of some money, went down to the market to sell a bracelet from the jewellery. There the merchants, who knew how to recognize true gems, looked at it and said:

'Why! this sham trinket has been made by someone with

knowledge of what he was doing, for these are pieces of glass and quartz coloured over with different colours and set in brass; there are no jewels or gold here.'

When the priest heard that he went off highly agitated, and bringing the store from his house he showed them the whole casket. They examined them again as before and pronounced them also to be all fakes; whereupon the fool appeared to be thunderstruck and went off straightway to argue with Shiva:

'Take your ornaments back and return me my own money.'

Shiva, however, said to him:

'How could I have any money now, for I have spent the whole amount in due course on my household?'

Then the priest and Shiva started quarrelling and so the two of them with Mādhava at their side approached the king. The priest gave the king the following account:

'All my riches have been consumed by Shiva for I did not know that this huge supposed treasure was made of skilfully coloured pieces of glass and quartz set in brass.'

Then Shiva said:

'O king, I have been a hermit from my childhood and, being entreated by this man, I was induced to accept payment and on that occasion I said to him that, in my ignorance of these things, he would have to judge on behalf of me who knew nothing about jewels and such things. He agreed to this, saying that he would guarantee it for me, so I, having received the whole inheritance, entrusted it to him. He then took it from me, O lord, for a price which he thought fit and here is the receipt which was signed by his own hand. From this the king knows now my great need for assistance.'

When Shiva's speech had come to an end, Mādhava then spoke:

'Do not look at the matter like this; you are worthy of respect; but what fault of mine is there here? For I did not receive anything from either you or Shiva. There was an inheritance from my people which lay for a long time in some other place and then it was brought here by me and I gave it to the Brahman.

If it is true that there is neither gold nor jewels there, then let it be said that I derived benefit from giving away glass and quartz set in brass. For my honest intention was clearly seen in that I gave with a sincere heart and have recovered from what was a very serious illness.'

After Mādhava had said this with a completely straight face, the king and all his courtiers laughed and were delighted with him. The general opinion of the assembly who were inwardly laughing was that neither Mādhava nor Shiva had done anything illegal, whereupon the priest who had lost his possessions went from there much embarrassed; for does not blind devotion to sheer greed always cause misfortune?

## The Fickleness of Fortune

THERE was once here on earth a city where there lived a king called Lakshadatta, the very foremost of patrons. He was incapable of giving less than a thousand coins to whomsoever asked him, while to those with whom he spoke he would give five thousand, so that he removed poverty from those with whom he was pleased, for which very reason he had been named Lakshadatta, the giver of riches.

Now there stood at the principal gate, day and night, a follower of his named Labdhadatta, dressed only in a tattered piece of skin for a loin-cloth. His hair was matted and in winter or in summer he would not leave his place even for a moment; the king observed him, but even though he was generous and compassionate, he never gave anything to him as he stood there, existing in endless misery.

Then one day that king went into a forest where game was found, and this attendant went on in front of him, carrying a staff. So, while the king with his retinue, seated on an elephant and carrying a bow, slew with his showering arrows tigers, boars and antelopes, his attendant going on alone in front on foot killed many boars and antelopes with his staff. When the king saw his strength, he thought within himself: Strange that this fellow should be such a hero! But he still did not give him anything. After finishing his hunting the king leisurely entered his city again, and his follower stood at the main gate as before. Then on one occasion King Lakshadatta set forth to subdue a kinsman of his who was a border chieftain, and a mighty battle ensued. In the battle his follower stood in front of him and laid low many an enemy with blows from the end of his stout wooden staff. When he had conquered his foe, the king returned to his own city, but even though he had observed his bravery he did not give him anything. In this way five years passed while the follower Labdhadatta stayed there living in distress.

When the sixth year came, King Lakshadatta noticed him by some working of fate, and, feeling compassion, he reflected: Up to this day I have never given this ever unfortunate fellow anything, so I do not see why I should not give him something and see whether or not the poor fellow's sins are forgiven him yet and whether good fortune will yet grant him a sight of her.

With these thoughts the king of his own accord entered his treasury and there filled a citron with jewels, as if it were a casket. He then called together a grand assembly, holding the meeting out of doors. Everybody, citizens, chieftains and ministers came along there. The king then called out in a kindly way to his follower who had entered among them:

'Come up here beside me.'

When Labdhadatta heard this he was delighted and when he had duly approached he sat down before the king. Then the king said to him:

'Recite something which you have composed yourself!'

When he heard this the follower recited the following words:

'"Just as the meeting of the rivers fills the ocean full, so does fortune flow to the rich, but she never comes within sight of the poor."'

When the king heard this he was very pleased and made him recite it once more, giving him the citron fruit full of real jewels. All the onlookers in that assembly were dismayed, however, and spoke freely amongst themselves, not knowing the truth of the matter, saying:

'Whomever the king is pleased with, he cuts away their poverty, but this follower is to be pitied, for the king summoned him with all courtesy and was pleased with him, yet gave him a citron; truly the wishing tree often turns out to be barren for the unlucky.'

The follower, however, took the citron and went away despondent, whereupon a religious mendicant named Rajavandin came up to him, and, when he saw the magnificent citron, he took it from him, giving him a garment instead. Then the mendicant went along and offered the fruit to the king, and the king, when he recognized it, said to the mendicant:

'Where did you obtain this citron, reverend sir?' Whereupon he explained to him that the attendant had given it to him.

At this the king became dismayed and astonished and he reflected: Alas! his fault is not even yet atoned for. So he took the citron and straightway leaving the assembly he carried out the duties for the day. The follower, however, went to his accustomed place at the main gate after selling the garment in exchange for food, drink and other things.

On the next day the king again held an assembly as before to which there came once more all the inhabitants of the town. When he saw the attendant entering there, the king again summoned him and made him approach. Once more he made him recite his stanza and being pleased with it he gave him the citron again in which the jewels were hidden. All the people said in astonishment at this:

'Look, this is the second day that the king has shown himself pleased with him to no purpose. What does this mean?'

The follower, though worried, took the fruit into his hand and, thinking that the king's pleasure was worthless, went away. Just then a certain official came up to him on his way to join the assembly in the expectation of seeing the king. When he saw the citron he took a liking to it and regarding it as a lucky omen he took it from the follower, giving him a pair of garments in exchange. This fellow then entered the king's assembly and, bowing at the feet of the king, gave him the citron as well as another present of his own. When the king recognized the fruit he asked the official:

'Where did you get this?'

'From the attendant,' replied the official.

Alas! Fortune does not yet grant him a sight of herself, the king thought to himself and he became exceedingly depressed. He took the citron and left the assembly. The follower, for his part, took the pair of garments and went to the market. He ate and drank after selling one garment and by cutting the other in half he made himself a pair of garments.

Then on the third day again the king called together an assembly and as before all the people attended it. Once more summoning the attendant and making him recite his stanza, the king again gave him the citron. At this the people were astonished, but the attendant went out and gave that object full of jewels to the king's wife. She, lithe as a tendril on the tree of the king's affection, gave him gold, as a flower might herald the fruit to come. When he bartered it the follower was in contentment that day. The king's wife, however, went into the presence of the king and she offered him that most delicious citron. When he recognized it once more he asked her where it had come from, whereupon she told him:

'It was given to me by the attendant.'

Then the king thought to himself: Even yet fortune does not look upon him; he must be of little merit not to know that my fruit is not without merit, for, lo! these precious jewels return to me again and again.

With these thoughts he took the citron and, putting it away safely, he rose up from the assembly and carried out the rites for the day.

On the fourth day the king once more held the assembly and it was filled with all the chieftains and attendants. Again the king had the follower summoned before him and after he had bowed down made him recite his piece. He further gave him the citron, and as it was being quickly handed over, it was only half caught in his hand and fell to the ground breaking in two. Then the many priceless jewels came out from where it had been broken at the join, and lit up the assembly-place. When they saw this everyone said:

'Indeed we did not know the true state of affairs, and lo! we have mistaken the matter in our confusion, yet such is the king's grace.'

When the king heard this he said:

'By means of this trick I tried to find out whether good fortune would really show herself to this man or not. But though his sins had not come to an end within three days, yet on the fourth they have, so that now indeed good fortune has let him look upon her.'

With these words, the king made that follower a border chieftain, giving him villages, elephants, horses and gold as well as those jewels. Then he rose from the assembly, with the people praising him, and went to perform his religious duties, while the follower went, with his desires fulfilled, to his own abode.

## Nāgasvāmin and the Witches

THERE was once a Brahman, Gomukha by name, who was
captured by the enemy in battle and, though he was taken
prisoner, he was abandoned by them a short time after in
the forest. There, in his distress, he determined to put an end to
his life by casting himself down a ravine, when a certain ascetic
came up to him and saved him by saying:

'Do not do this, Gomukha, for you will once again behold
your king when he has become victorious.'

When Gomukha then asked him: 'Who are you? How do
you know this?' the latter answered: 'Come to my hermitage!
I will explain to you there.' Upon this Gomukha went with
him to his hermitage, an abode of Shiva, for the abundance of
his learning had been proved by his knowing Gomukha's name.
Once there, the ascetic performed the rites of hospitality and
then related his story as follows:

I am a Brahman, Nāgasvāmin by name, from the city known
as Kundina. When my father went to heaven I myself went to
sit at the feet of the teacher Jayadatta for the sake of acquiring
knowledge. Yet, though I was instructed, I never learnt a single
syllable on account of my dullness, for which reason all the

other pupils there used to laugh at me. Then, being consumed
by this contempt, I set out to worship the goddess Chandī who
dwells in the Vindhya mountains. Yet but half-way there I
reached a city where, as I entered for the purpose of begging
alms, a lady from one of the houses gave me a red lotus together with alms. I took it but when I reached another house,
and the mistress there saw me, she said to me:

'Woe to you! You have been possessed by a witch. Look! In
the guise of a red lotus you have been given a man's hand by
her.'

At these words I looked, and indeed it was a hand and not a
lotus. I let it go and then, falling at her feet, I said to her:

'O mother, show me some way out of this, that I may yet
live, for this misfortune could cause my death.'

When she heard this, she said to me:

'Go then, there is in the village of Karabha, three leagues
distant from here, a Brahman known as Devarakshita. He has
in his house there a magnificent brown cow, like Surabhi herself, the sacred cow of the gods, and she will watch over you
this night if you reach her refuge.'

When she had said this to me, I ran in my fear and by the
close of day I reached the house of that Brahman in the village
of Karabha. When I entered there I saw the brown cow and I
bowed before her, telling her:

'In my fear I have sought your help, O goddess.'

Just at that moment the witch together with the others came
there by night through the sky, threatening me and thirsting
for my flesh and blood. When the brown cow saw this, she
placed me between her hooves and so protected me, fighting
with these witches the whole night long. In the morning they
disappeared, and then the cow spoke to me with a clear voice:

'My son, I shall not be able to protect you from now on;
therefore go, for in a forest five leagues from here there dwells a
most learned ascetic, the excellent Bhūtishiva. He will protect
you this night if you seek refuge with him.'

When I heard this, I bowed down before her and then I set

out and having swiftly sought out that Bhūtishiva, I placed myself in his care. Once again in the night the witches came as before, but Bhūtishiva made me enter his house and with his trident in his hand he stood at the door for the whole night and menaced the witches. When he had overcome them, Bhūtishiva gave me food the next morning and told me:

'Brahman, I shall not be able to protect you again, but there is in a village ten leagues distant from here a Brahman named Vasumati; go to where he is and then when you have survived the third night you will be freed.'

At these words of his, I bowed to him and then set forth. On account of the distance, however, the sun went down while I was still travelling on my journey, so that the witches came up behind me in the night and laid hold of me. And when they had seized me they rose up delighted through the sky. At that moment though some other witches from an unknown quarter flew past in front of them, and unexpectedly there arose a disordered battle between them all, whereupon I fell from their hands into a deserted spot.

Then I saw a great and solitary mansion which seemed to invite me with its open door as though saying: 'Come in!'

I fled in there and, when I had entered it, bewildered with fear, I saw a woman of miraculous loveliness surrounded by a hundred other women. She was radiant as some healing and protective flower made by the creator in his compassion for me, lighting the darkness, as it were, with her lustre. As soon as she was questioned by me she comforted me and answered me:

'I am a goddess, Sumitrā by name, dwelling in this place on account of a curse. It has been decreed for the atonement of the curse that I should marry a mortal, so you who have come here by chance should take me to wife, nor have fear on this account.'

When she had said this she speedily commanded her attendants and honoured me to my delight with baths and soothing ointments and clothes as well as food and drink. No comparison could there be between the terror from the witches and the joy

of that moment! Truly the coming of joy or sorrow is not under-
stood even by fate.

There I passed the days in happiness with that goddess, and
then one day she said to me of her own accord:

'The curse is today at an end for me and so now I depart
from here, yet through my favour a divine intelligence will be
yours to have; and you will become a religious ascetic in perfect enjoyment and without fear, but if you remain in this house
of mine you must never look at the middle room!'

When she had said this she disappeared, and then out of
curiosity I climbed to the middle storey and there I saw a horse.
When I got to the horse I was struck a blow from his hoof and
at that instant I found myself standing in this abode of Shiva.
From that time on I have remained here and I have gradually
developed supernatural faculties so that in this way, even
though I am a mortal, I have come to know the past, the present
and the future. In the same way too, everyone can gain the
accomplishment of all his desires though he may be beset with
misfortune. Remain here therefore! Shiva will grant you the
fulfilment of all your desires!

When Gomukha had been told this tale by that wise man,
he remained some days there in the hermitage, living in the
hope that he would worship at the feet of his king once more.
Then Shiva related to him in a dream of his lord's victory, and
straightway he was taken by some heavenly nymph and led
back to the king.

## Love at First Sight

ONCE upon a time King Pālaka of Ujjayinī instituted a
spring festival. When it had begun and the people were
engrossed in it and setting up a confusion of shouting, a
wild elephant which had broken its fastenings ran in unexpectedly amongst them. The elephant made short work of the
driver's hook and tossed the driver himself from its back; then,
wandering about inside the city, it rapidly killed very many
people. The elephant keepers ran after it as well as the citizens,
but nobody was able to hold the elephant in check. Gradually,
as the elephant wandered about, it reached a Chandāla settlement, where only people of the lowest caste lived. Then, from
one of the huts there, a Chandāla maiden came out. She made
the ground radiant wherever the lotus-like beauty of her foot
stepped and was as if pleased that the moon, her rival, had been
overcome by her face; she gave peace to the eyes of everyone as
they stood stock still and as if asleep before her, for their minds
were turned away from other things.

The girl went up face to face with that mighty elephant and,
striking its trunk with her hand, she scolded it with arched side-
long glances. The elephant was infatuated by the touch of her
hand and, bending its head down, it was pierced by the sight

of her as it gazed upon her, nor did it move one step from there. Then the lovely girl amused herself by climbing on to a swing which she had made from her outer garment and had flung over the elephant's tusks. And when it saw that she was distressed by the heat, the elephant turned into the shade of a tree.

When they saw this great wonder, the citizens there all said: 'Lo! truly this maiden must be divine, for even the animals are taken by her magical power which like her beauty excels everything.'

In the meantime the Prince Vardhana had come to know of the tumult and, approaching out of curiosity, he saw the maiden. And as the prince gazed on her, the fleet deer of his thoughts was bound in its flight by that net of the hunter of love. She too, noticed him and her heart being taken by his handsomeness, she descended from her swing on the elephant's tusks and took her outer garment. Then, while a driver mounted the elephant, she went into her own house, ever looking back at the prince in modesty and love.

Vardhana, however, now that the disturbance caused by the elephant was over, went off despondent to his palace, his heart quite taken by the girl.

There, as he was tormented by not having that lovely maiden, and forgetting that the spring festival had begun, he said to his friends: 'Do you know whose daughter she is? And what is the girl's name?'

When they heard this, his friends told the prince: 'There is here in a Chandāla settlement, a certain man named Utpala-hasta and he has a daughter Rati by name, but for high caste people her form is fair in so far that there is delight in looking at it, as if it were painted in a picture and were not to be embraced.'

Hearing this news from his friends, the prince said to them: 'I do not consider her to be the daughter of an outcast; she is certainly some heavenly maiden, and if I do not obtain her with her beauty for my wife, what is the point in my living?' So the prince spoke with his friends nor could he be dissuaded for he was entirely consumed with the fire of separation from her.

Thereupon his parents, when they came to learn of it, were for a long time in despair.

'How can our son, born in the lineage of kings, desire a maiden of another caste?' said the queen to King Pālaka and he answered her: 'This girl must be an outcast only through some trick of fate; in reality she must be some other maiden since the heart of our son is set on her; for the minds of the good, either by being attracted or by being loath, tell the difference between what may be done and what must be avoided. If you have not heard the following tale, my queen, please listen to it now.

'There was formerly in a city known as Supratishthita, a King Prasenajit who had a very beautiful daughter, Kurangī by name. One day she went out into the gardens and was flung up together with her couch on the tusks of an elephant which had broken its fetters. As her attendants ran away she cried out and thereupon a certain young Chandāla seized a sword and ran towards the elephant. Cutting at its trunk, he killed the elephant with a blow of his sword and so the hero rescued the king's daughter. Then, gathering her retinue together again, she went to her own palace, her heart taken by the wealth of his beauty and bravery; and oppressed by separation from him she remained there thinking: Either he who rescued me from the elephant must be my husband, or else death must be.

'The young Chandāla, for his part, went to his own house and with his mind smitten by her loveliness, he suffered torment as he thought about her:

'What comparison is there between me, a man of the lowest caste, and her, the daughter of a king? And how indeed could there be such a marriage between the royal swan and a crow? I cannot either speak or even broach such a ridiculous subject. Therefore death is my only refuge in this terrible strait.

'With these thoughts he went to the grove of the departed

ancestors and, after bathing, he made a funeral pyre and setting it ablaze he declared to Agni, god of fire:

"'O god, purifier, soul of the universe, since I am giving myself as an offering to you, may that princess be my bride in a future birth."

'When he had said this and was about to cast himself into the fire, Agni, being pleased, made himself manifest in visible form and spoke to him:

"'Do not act hastily! She will indeed become your wife, for you were not a Chandāla always; as for what you were, hear that from me now. There dwells in this city an excellent Brahman called Kapilasharman and in the hearth of his sacred fire in actual bodily form, I dwell. There one time as his daughter came near me, I, being smitten with her beauty, took to wife that girl, whose offence was dispelled by my charm. Then, through my power, you were born to her, my son, but you were straightway cast by her in her shame on to the open highway. Then some Chandālas picked you up and nourished you on goat's milk. So you are my own son born of a Brahman woman and therefore there is no impurity in you. You have arisen out of my lustre and you will obtain the king's daughter Kurangī as your wife."

'When he said this, the fire god disappeared, while that adopted son of an outcast, in a state of extreme happiness, took the matter into consideration and returned to his own home. Then King Prasenajit, being inspired in a dream by Agni, gained insight into the truth of the affair, and gave his daughter to that son of the purifying god.

'In this way,' King Pālaka went on, 'there are always, my queen, divine beings on earth and this girl Rati is also some divine creature and is not of the lowest caste. For certainly such a jewel is different, and undoubtedly she was the beloved of my son in some other birth as has been shown by this love at sight.'

Thereupon King Pālaka sent messengers to ask Utpalahasta for his daughter. When the outcast was requested by the messengers, he said to them: 'My wish is that my daughter Rati will be given only to him who shall cause the eighteen thousand Brahmans dwelling in the town to eat in my house.'

When the messengers heard these words of his that contained a promise, they returned and informed King Pālaka about it.

When he pondered that there was some reason in this, the king assembled the Brahmans in the city of Ujjayinī and, telling them the story, said to them:

'The eighteen thousand of you must eat in the house of an outcast here, Utpalahasta; none other is my wish!'

Being thus addressed by the king, the Brahmans, who were afraid of the Chandāla's food and in bewilderment as to what to do, assembled at the shrine and performed penance to Shiva the avenger. Thereupon they were commanded in their dreams by the god Shiva:

'Eat food in the house of the outcast Utpalahasta without fear, for he is a *vidyādhara*, a demigod, and has no family ties with the Chandālas.'

At this they arose and going to the king they told him about it and then went on to say:

'Let Utpalahasta cook clean food somewhere other than at the Chandāla settlement, O king, and we will then eat it here.'

When the king heard this he was pleased and commanded another house for Utpalahasta, and made him cook food for him there with clean cooks. When then Utpalahasta had bathed and put on fresh clothes he stood before them, and the eighteen thousand of the highest birth had their meal.

When they had eaten, Utpalahasta approached the king in the presence of his people and bowing to him said:

'There was an eminent lord of the *vidyādharas*, the fairies who attend upon Shiva, Gaurīmunda by name, and I was a dependant of his. When my daughter Rati was born, O king, Gaurīmunda spoke to me in secret as follows: "That son of the king of the Vatsas is to be our future emperor, so the gods say; he is a thorn in our side for as long as he remains without the empire, so depart forthwith by your magical power and destroy him!"

'When I was sent in this way by the evil Gaurīmunda and was journeying through the sky on that account, I saw Shiva before me. The lord, being furious, made a terrible sound and swiftly cursed me: "How can you, you wretch, perform wickedness against a noble-minded person? Go with this very body of yours together with your wife and daughter and fly down amongst the Chandālas in Ujjayinī, you evil-minded fellow! When someone shall have fed eighteen thousand of the Brahmans dwelling in the city in your dwellings by way of purchase money for the gift of your daughter Rati, then your curse will come to an end, and your daughter shall be given to him who supplies that purchase money."

'When Shiva had said this he disappeared and I then, with the name of Utpalahasta, fell among the lowest caste, but I did not mix with them. Today now my curse is at an end, thanks to the favour of your son. Now therefore I go to my dwelling among the *vidyādharas* in order to pay my devotion to the emperor.'

On saying this, the Vidyādhara handed over his daughter and, flying up into the sky with his wife, he whom they had known as Utpalahasta left them.

King Pālaka, being delighted then on knowing the truth, gave orders for the wedding of his son and Rati. And his son Vardhana, now that he had obtained a *vidyādharī* as his wife, remained there with the attainment of his wishes fulfilled beyond all expectation.

## The Foolish Brahman

THERE was once a Brahman, Devasharman by name, in the town of Devakotta. One day someone gave him a present of a pot of barley, so, taking it with him, he went to a potter's shed which was filled with pots of every size; there he lay down and there he began to ponder during the night:

If I were to sell this pot of barley I would get ten gold pieces; with that money I could buy pitchers and pots. These I could sell at a profit and, with my money increased, I could trade in *betel* nut, silks and the like, so that soon I would be worth thousands. This way I could marry four wives, and of course I would especially favour the most beautiful among the four. Then, if her co-wives began to quarrel with her out of jealousy, I would get angry and beat them with a stick like this!

Hereupon he lashed out with a stick he was holding in his hand. As a result, not only was his own pot of barley smashed, but many other pots as well.

When the potter, who had been roused by the sound of the pots being smashed, saw what had happened, he swore at the Brahman and drove him out of his shed. Truly it is said that he who rejoices about things which have not yet happened is likely to suffer great scorn.

## The Vetāla's Stories

THERE is a place on the banks of the Godāvarī where once upon a time there lived a well-renowned king, known as Trivikrama, and his might was as that of Indra, the chief of the gods. Every day a religious mendicant, Kshāntishīla by name, used to approach the king in his assembly hall in order to pay him respect, and used to give him a fruit. The king, for his part, after taking the fruit, would hand it every time to the keeper of his treasury who stood beside him.

In this way ten years went by, and then one day as the mendicant gave the king the fruit and left the assembly hall, the king gave it to a young pet monkey that had entered there by chance, having escaped from the hands of its keepers. As the monkey set about eating the fruit, there fell from it, as it split in the middle, a rare and priceless jewel. When the king saw it and picked it up, he asked the keeper of his treasure:

'Where are you in the habit of putting the fruits which that mendicant brings and which I am constantly handing to you?'

When he heard this the treasurer anxiously informed him:

'I threw them through a window into a store-room without opening it up; if you will command me, then, my master, I will open it up and investigate.'

When he had said this the treasurer was given permission by the king and so he went straightway and, on his return, he informed his lord:

'I have seen those fruits withered away in the treasury and now I see a heap of jewels like a profusion of blazing rays, O king!'

On hearing this the king was pleased and gave these jewels to the treasurer, while on the next day he questioned the mendicant who came as before:

'Mendicant, why do you honour me every day with such an extravagance of gems? I will not take this fruit from you now unless you tell me.'

Then the mendicant said secretly in answer to the king's words:

'I have a rite to perform which requires the assistance of a brave man, and that, O great hero, is why I beg you to afford me aid.'

The king readily assented to what he had asked, and then the ascetic was pleased and went on to say:

'Then you must come to me in the approaching dark fortnight, when the moon turns from full to new, at the onset of night. I shall stay in wait at the foot of a *banyan* tree at the bottom of the great cemetery.' When the king had agreed, saying: 'Very good, I will do this,' the mendicant Kshāntishīla was delighted and went to his own abode.

Then the noble-minded king, when the next dark fortnight had come, remembered the promise which he had made to that mendicant in his successful entreaty and so, with his head adorned with a dark cloth, he went out from the palace, sword in hand and unperceived. And he set out undaunted for the cemetery which was murky with a cloud of thick and terrible darkness, a cruel night with the flames from the fearful burning of the funeral pyres. It was hideous with the skulls and bones and skeletons of the corpses just visible, and alive with the horrid ghosts and *vetālas* (or demons) who delightedly gathered there, a place of deep dread like another form of Shiva in his

role as lord of terror, thundering with the great howlings of
jackals. He made out where the mendicant was and found him
making the design of a circle under the *banyan* tree, whereupon he approached him and said:

'Here I have come, O mendicant, tell me what I may do for
you.'

When he heard this the mendicant looked up and delightedly
spoke to the king:

'O king, if you will do a favour, then go south from here and
you will find a solitary blackwood tree; there is a man there,
hanging dead from it; go and bring him here; give me this
assistance, O hero.'

When he heard this, the king, true to his promise, said that
he would, and turning to the southern part, the brave man went
off. There, as he went along a track lit by the light of the burning pyres, he somehow reached the blackwood tree in the darkness. And there, on a main branch of the tree which was
scorched by the smoke of the pyres and smelt of gore, he saw a
corpse hanging as if on the shoulder of a ghost. Thereupon he
climbed the tree and by cutting the rope let the corpse fall to
the ground, and when it fell it unexpectedly cried out as if it
had been hurt. Then the king climbed down and thinking it
might be alive he rubbed its limbs out of compassion, whereupon the corpse uttered a raucous laugh. Then the king realized
that it was possessed by a *vetāla* and just as he was unper-
turbedly saying: 'Why are you laughing? Come, let us go!' he
no longer saw the corpse with the *vetāla* on the ground, but
he perceived it then once more hanging from the tree. Then he
climbed up once more and let it drop; for the diamond-like
hearts of the brave are less breakable than adamant. And in
silence then, King Trivikrama put the corpse with its *vetāla*
up on his shoulder and began to set forth. As he was going
along, the *vetāla* dwelling inside the corpse on his shoulder
said to him; 'O king, I shall tell you a tale to divert you on the
way, so please listen!'

## SOMAPRABHĀ AND THE THREE SUITORS

There was a Brahman in Ujjayinī, a favourite dependent of
the King Punyasena, a minister with virtuous qualities, whose
name was Harisvāmin. And to that householder's wife, who
was his equal in birth, there was born an equally virtuous son
whom they called Devasvāmin. And a daughter who was renowned for her beauty, a loveliness like none other, was also
born to him, and her they rightly named Somaprabhā, which
means Moonlight. When the time came for this daughter to be
given in marriage, being conceited about her extraordinary
beauty, she said, through her mother, to her father and brother:

'I must be given in marriage to a hero, or to a man who
possesses knowledge, or else to one who has magic power; I
must not be given to anyone else if there is to be any purpose
in my life.'

When he heard this, her father Harisvāmin kept thinking
about how he might gather for her a husband of one or other
kind, when he was sent with a mission by the King Punyasena
for the purpose of concluding peace with a king who had come
from the Deccan to make war. And when he had carried out
this duty, a most excellent Brahman came to him, on hearing about the abundance of her beauty for the hand of his
daughter.

'My daughter wishes for no other husband than a man who possesses knowledge or magic powers or who is a hero; tell me then, which of these are you?'

When Harisvāmin had addressed him thus, the Brahman who sought his daughter answered him: 'I possess magic powers.' When then he was further pressed to demonstrate them, the magician conjured up a sky-going chariot by his own power. Straightway then he made Harisvāmin mount the supernaturally contrived chariot, and conducting it he showed him the worlds with their various heavens. And then he brought him back delighted to that very camp of the king from the Deccan where he had come on account of his embàssy. So Harisvāmin promised his daughter to the magician and settled the marriage for seven days' time.

At that same time in Ujjayinī, his son Devasvāmin was visited by another Brahman who had come for his sister. And when he was told that she would have none other for a husband than one who possessed knowledge or magical power or who was a hero, he declared himself to be a hero. When he had then displayed his skill in weapons and missiles, Devasvāmin promised to give that hero his sister. And he too, according to the reckoning of the astrologers, fixed his wedding for the seventh day, having made the decision without his mother knowing.

But at the same time also, his mother, Harisvāmin's wife, was asked separately by some third man for the hand of her daughter. When he was told by her: 'Our daughter's husband must prove to be one who either possesses knowledge or is a hero or a magician,' he answered her, saying: 'Mother, I possess knowledge.'

On questioning him about the past and the future, she too promised to give her daughter to that knowledgeable fellow on the very same seventh day.

On the next day then, Harisvāmin returned and related to his wife and son the decision which he had taken regarding the giving of his daughter in marriage. And the two then told him,

each one separately, what they had done themselves, whereupon he became much distressed on account of three bridegrooms having been invited.

Then on the day of the wedding those three bridegrooms, the
one who was knowledgeable, the one who possessed magical
power and the hero, came to the house of Harisvāmin. And at
that very moment the girl Somaprabhā strangely and un-
accountably disappeared somewhere and although they looked
for her they could not find her. Then, in his agitation, Harisvāmin said to the one who possessed knowledge:

'You with knowledge, tell me now quickly where has my
daughter gone?'

At these words the knowledgeable fellow said:

'She has been carried off by a *rākshasa*, the demon Dhūmra-
shikha, and taken to his own abode in the Vindhya forest.'

When he was told this by the possessor of knowledge,
Harisvāmin said in his fear:

'Ah alas! How can she be rescued, how can there be a
wedding?'

When the possessor of magical power heard this he spoke up:

'Take courage! I shall lead you there this very moment,
to the place where the possessor of knowledge has said
she is.'

With these words he created a chariot as before, equipped
with all kinds of weapons, and making Harisvāmin as well as
the possessor of knowledge and the hero climb into the aerial
car, he straightway conducted them to the dwelling of the
*rākshasa* which had been described by the knowledgeable
fellow to be in the region of the Vindhya forest. There the hero,
on behalf of Harisvāmin, battled with the angry *rākshasa* who
came forward on finding what had happened. Then there took
place between those two, the hero and the *rākshasa*, a miraculous fight, like that of Rāma and Rāvana who fought for a
woman's sake with arms of every kind. And soon the hero had
cut off the head of the *rākshasa*, even though he was hard to
overcome in battle, with a crescent-headed arrow. Once the

*rākshasa* was killed, they all returned in the chariot of the possessor of magic, together with Somaprabhā who had been rescued from the demon's dwelling.

When they reached Harisvāmin's house, and as the auspicious moment was about to take place, there arose a great quarrel between the possessor of knowledge, the possessor of magical power and the hero. The possessor of knowledge said:

'If I had not known, then how could the maiden, who was hidden away, have been rescued? So she should be given to me.'

The possessor of magical power maintained, however:

'If I had not made you a chariot to travel the sky, how could you have gone and come back as quickly as if you were gods? And how could there have been a battle with the *rākshasa* who also possessed a chariot, if you had not had a chariot? Therefore she should be given to me, for the auspicious moment has been gained by me.'

The hero, however, said:

'If I had not killed that *rākshasa* in battle, then in spite of the effort which you two made, how would the girl have been brought back? Therefore she should be given to me.'

And as they went on quarrelling like this, Harisvāmin remained there silent, greatly disturbed in his mind.

'To whom indeed should she be given? Your majesty, tell me this; for if you know, and do not tell me, your head shall split in pieces.'

When King Trivikrama heard this from the *vetāla*, he broke his silence and said to him:

'She should be given to the hero, through the might of whose arm she was won, owing to his staking his life when he slew that *rākshasa* in battle. The possessors of knowledge and magical power were put there by the creator, however, in the role of assistants; for are not soothsayers and artificers always in the service of others?'

When he heard this pronouncement from the king, the

*vetāla* straightaway went from on top of his shoulder back to his own place; whereupon the king once more unhurriedly set out after him.

## ANANGARATI AND THE MEN OF FOUR CASTES

When King Trivikrama returned once more, he took the *vetāla* on to his shoulder from the blackwood tree and as he set out he was addressed by the *vetāla*:

'Your majesty, what can your kingdom have in common with wandering around in this cemetery at night? Do you not see that this ancestral grove is beset with ghosts, fearful with the night and closed as it were with the thick smoke from the funeral pyres? Alas! What kind of perseverance is this, all for the sake of helping this mendicant? Anyway just listen to this riddle of mine to divert you on the journey.'

There is in the land of Avanti a city built by the gods at the beginning of the ages, like the body of Shiva, proud and adorned with enjoyment and riches. In it there dwelt a king called Vīradeva, the best of monarchs, who had a chief queen whose name was Padmarati.

One day then the king went with her to the banks of the River Mandākinī and out of his desire for a child he pleased the

god Shiva by performing penances. When he had remained for
a long time there and observed the rites of bathing and devotion, he heard a voice from heaven, from the god who was
pleased.

'O king, a son will be born, a hero to carry on your lineage
and a daughter who will surpass the heavenly nymphs by a
beauty unlike any other.'

When the King Vīradeva heard this voice from heaven and
that his wishes were gained, he returned to his own city together with his queen.

There to the Queen Padmarati was born as the first-born, a
son named Suradeva and in course of time a daughter, and her
father named her Anangarati, saying that she would arouse
love even in Ananga, the bodiless Kāma, god of love. When she
reached marriageable age, her father sought to find her a
suitable husband and he had brought to him all the kings on
earth, painted on canvas. But as not even one of them appeared
to him to be her equal, out of his affection for her the king said
to his daughter:

'I just do not see a fitting suitor for you, my daughter, so
assemble all the kings together and perform your *svayamvara*,
choosing your husband for yourself.'

When she heard her father say this, the princess spoke:

'My dear father, I would be too shy to appear at a *svayamvara*, but you must give me in marriage to some young man of
good appearance who knows one art thoroughly; I do not want
anyone better than that.'

When the king heard this resolution from his daughter
Anangarati, he sought around for a husband of that kind for
her. Then four men who were heroes, wise and handsome, and
who had heard about it from rumour, came to visit him from
the Deccan. These suitors were honoured by the king and in the
presence of his princess they each related to him their own
particular skill.

The first one said: 'I am of Shūdra caste, Pancaphuttika by
name, and every day I make by myself five pairs of the finest

garments. Of these I give one to the god; one to a Brahman; one I retain as clothes for myself; one I would give to my wife, if this lady here were to be mine, and one I sell and so provide food and drink. Let Anangarati be given to me since I possess such skill.'

When he had finished speaking the second man said: 'I am of the Vaishya caste, Bhāshājna by name, and I understand the speech of every bird and animal; therefore let this princess be given to me.'

When the second man had spoken, the third man then declared: 'I am of the Kshatriya caste, a strong-armed king, Khadgadhara by name, and there is not a match on earth for me in knowledge of the science of sword-play; therefore, O king, you should hand your daughter over to me.'

When the third man had spoken, the fourth had this to say: 'I am of Brahman caste, by name Jīvadatta, and my knowledge is such that I can even take creatures who are dead and show them quickly to be alive; so let this maiden be given to me who am famous for having accomplished such glorious deeds.'

King Vīradeva then, to do justice to his daughter, looked at those men who had spoken thus and who were handsome and dressed like gods, and he appeared as if assailed by doubt.

When the *vetāla* had related this story, he then questioned the king, frightening him with the curse described before:

'Now may your highness tell me, to which of these four men should the girl Anangarati be given, O lord of men?'

When he heard this, the king then answered the *vetāla*:

'You are making me break silence mostly in order to waste time, otherwise, how could this question, you master of magic, be anything but a reproach? How could a girl of Kshatriya caste be given to a Shūdra weaver? For that matter how could a girl of Kshatriya caste be given to a Vaishya? And as for the knowledge of the language of beasts and so on which he possesses, how can this be used in trade? And as for the third man here, the Brahman, he must be fallen from his rank and have swerved from his religious duties in that he considers himself a hero and possesses magical tricks. Therefore she must be given to the fourth man, the Kshatriya, Khadgadhara, who is of the same caste and who is renowned for his knowledge and heroism.'

When the *vetāla* heard this answer from the king, he disappeared speedily from his shoulder as before, back to his own place, by his magic power. The king, for his part, once more set off to bring him back; for weariness does not gain entry into the heart of a hero which is cased in the armour of fortitude.

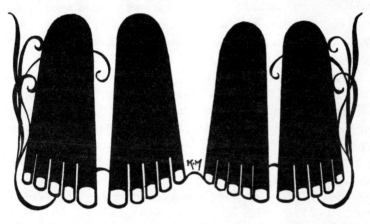

CHANDASIMHA, HIS SON AND THEIR TWO WIVES

Then that heroic King Trivikrama, taking no account of the demon-filled night, clouded with darkness lit up by the eyes of the funeral pyres in that fearful cemetery, went once again to the blackwood tree and took the *vetāla* down from it. And so he put it on his shoulder and as he set forth as before, once more the *vetāla* spoke to that lord of men:

'Great king, I am distressed, even if you are not, with these comings and goings, so now just listen while I relate this most difficult question.'

There was a certain king, ruler of a country in the Deccan, who had many relations and was foremost of worthy men, by name Dharma. He had a wife called Chandravatī from the country of Mālava, born of noble lineage, a veritable adornment of virtuous womanhood. And by that wife the king had but one child, a daughter named Lāvanyavatī.

When the time came for the daughter to be given in marriage, King Dharma was uprooted by his kinsmen who had joined forces and overturned the kingdom. Then, taking flight, he left his country together with his wife and daughter in the night, taking with him a store of precious gems. He determined then to set out for Mālava, the land of his father-in-law, and by night-time he reached the Vindhya forest, accompanied by his wife and daughter. And as that king entered it, the night, as if shedding tears with its frosty drops of dew, left him after having paid him attendance, as it were.

Then the sun, climbing up the eastern mountain, spread out its rays in front as if to ward him off, saying: 'Do not go into this brigand's forest!' As the king continued then on foot with his wife and daughter, their feet hurt by the points of the *kusha* grass, he reached a small village of the Bhillas, full of men accustomed to taking the lives and property of others and void of worthy people.

When they saw the king coming at a distance, with his ornaments and fine clothes, numerous Bhillas, armed with all kinds of weapons, ran forward to plunder him. When he saw them, King Dharma said to his wife and daughter: 'The barbarians will set upon you first, so go into the forest.'

At this command from the king, the Queen Chandravatī entered the forest together with her daughter Lāvanyavatī. The brave king, however, armed with sword and shield, slew many of the Bhillas who came against him, raining showers of arrows. Then the whole village was summoned by their chief, and they fell upon the king who stood alone, breaking his shield with their blows, and killed him. The hordes of heathens took his ornaments and went off, while the Queen Chandravatī, who had stayed some way off in a thicket in the forest, after seeing her husband slain, fled in distress with her daughter and entered

another dense and distant forest. There even the shadows had
drawn in by the cool roots of the trees as if afflicted, like way-
farers, by the noonday heat. In a spot on the bank of a pool, in
the shade of an *ashoka* tree, the queen, pained with sorrow,
weeping and exhausted, sat down with her daughter.

At that time, a certain man of high rank who lived near by
had gone to that forest on horseback, accompanied by his son.
His name was Chandasimha and when he saw the footprints
of the pair marked in the dust, he said to his son: 'Let us follow
up these lucky little signs, and if we catch up with these two
women, then you shall choose one of them for yourself as you
please.'

At these words his son replied: 'Whichever one has these
dainty feet would please me as a wife, for, I should imagine,
she would be the younger, and therefore suitable for me; but
the one with the big feet would be suitable for you, being
older.'

When he heard these words of his son, Chandasimha said to
him: 'What talk is this? Your mother has only just gone to
heaven; and when one loses such a good wife, how can one
think of another?'

'Father, it should not be so,' said his son to Chandasimha.
'For the house of a householder is an empty place without a
wife; and have you not heard the words which Mūladeva
wrote: "Where one's beloved with shapely breast and waist
does not sit awaiting one's return, what sensible person would
enter such a prison without bonds which passes for a home"?
So, father, you will have ill luck if you do not take to yourself
as wife the companion of the one that I have chosen.'

When he heard his son saying this, Chandasimha assented
and he continued with his son leisurely to follow the row of
footprints. And when they reached the place where the pool
was, he saw the Queen Chandravati, a dark beauty adorned
with a wealth of pearl necklaces, seated in the shade of the
tree, shining like the night sky in the noonday by her daughter
Lāvanyavatī as dazzling as the white moonlight. He eagerly

went up to her accompanied by his son, but when she saw him she rose up in terror thinking that he was a thief.

'Away with your fear! These two are not thieves, mother; assuredly these two with their handsome looks and fine clothes have come here on a hunting expedition.' At these words from her daughter the queen was uncertain, but Chandasimha got down from his horse and said to them both: 'Why such panic? We two have come here because we wanted to see you; so be at ease and tell us without fear who are you who have entered this truly uninhabited forest? For you are indeed worthy of dwelling in the finest of palaces. How have your feet, accustomed to the lap of the ladies-in-waiting, wandered on this thorny ground? To think of it is painful to our minds. Tell us the story of your adventures, for our hearts are pained, and we are not able to look upon you in a forest full of beasts of prey.'

When Chandasimha had said this, the queen slowly breathed a sigh and, worried by shame and grief, related to him what had happened to them. Then when Chandasimha realized that she had lost her husband he comforted her and her daughter and, winning them with gentle speech, he made them his own. Then he with his son placed her and her daughter on the backs of their horses and conducted them to his own magnificent mansion. And she for her part entrusted herself to his care, as if she had been born again, for what can a woman do who has lost her husband and fallen into misfortune in a foreign land?

Then, on account of the smallness of her feet, the son of Chandasimha declared Chandravatī to be his wife. And her daughter, the Princess Lāvanyavatī, was made his wife by Chandasimha, because her feet were the larger. For this had been agreed upon previously when they had seen the two sets of marks, the small and the large, made by the two women's feet; and who, indeed, ever denies his pledge?

In this way, on account of the mistake over the feet, since they had become the wives of the father and the son, the mother became her daughter's daughter-in-law and the daughter became her mother's mother-in-law. In course of time they and their husbands had sons and daughters, and these also had sons and daughters. And so Chandasimha and his son, having obtained their wives, Lāvanyavatī and Chandravatī, lived happily ever after.

When the *vetāla* had told this tale on the way during the night, he once more questioned the King Trivikrama.

'Now that you know, your majesty, that from the mother and daughter and the son and the father there came offspring on both sides in course of time, then tell me this: what were the relationships between them? And the curse I mentioned before will befall you if you know but do not answer.'

When the king heard this from the *vetāla*, although he turned it over many times in his mind, he did not know the answer and so he went on in silence.

Then the *vetāla* in the dead man's body, placed on the top of the king's shoulder, thought to himself, laughing inwardly: This king was not able to answer this great question, and so delighted he goes along with much swifter step. No longer can I deceive this most valiant king, nor will these attempts to gain time prevent that ascetic Kshāntishīla from playing his tricks with us. I will therefore deceive that evil-minded fellow and so bestow the success which he seeks upon this king whose future will be illustrious. With these thoughts in mind, the *vetāla* then said to the king:

'Your majesty, in spite of the weariness of these comings and goings in this cemetery terrible with the dark night, you appear as if content, nor are you in any indecision. I am pleased with this wonderful fortitude of yours. Take this corpse on now, for I am going to leave it; and now listen, for I am speaking for your good, so do this! He on whose behalf you have brought this corpse is a rogue of a mendicant; it is I whom he wishes to summon in this corpse in order to pay me devotion. He wishes to make a sacrifice of you, so when the wretch says: "Make a humble bow with your face to the ground!" then, your majesty, you must say to that mendicant: "Show me first just exactly how I am to perform this!" Then, as he falls to the ground to demonstrate the bow to you, cut off his head with your sword. When you have done that, you will gain what he has sought, namely, the succession to the lordship over the *vidyādharas*, the fairies who attend upon Shiva; meanwhile, enjoy your reign upon this earth by making a sacrifice of him. Otherwise the mendicant will offer you up as a sacrifice, which is why I have been hindering you for so long. So may you have success.' With these words the *vetāla* left the body of the dead man on the king's shoulder and went off.

Then the king, as a result of what the gratified *vetāla* had said, thought about that most unpleasant ascetic Kshāntishīla, but in good spirit he took up the body of the dead man and set out to meet him under the *banyan* tree.

Soon King Trivikrama saw the ascetic awaiting him at the foot of the tree in the cemetery made fearsome by the dark of the moonless night. He was standing in a blood-smeared circle marked out with the whitened powder of bones, and at the four cardinal points around the circle he had placed pitchers brimming with human blood. All was lit up splendidly with candles of human fat, and near by was a fire upon which offerings were being burnt. Everything was ready for the sacrifice, and the ascetic was deep in meditation upon the deity of his choice.

The king approached him, and the ascetic, seeing that he had brought the corpse, was delighted and came forward to greet him with flattering words.

'Your majesty, you have favoured me exceedingly by accomplishing this most difficult task; for one little expects to see a person of your importance busying himself thus in such a place and at such a time as this. Truly your excellence is unwavering; you are the foremost of kings.'

As he was saying this, the ascetic took the dead body down from the king's shoulder. He then bathed and anointed it and, casting a wreath around it, he placed it in the centre of the circle. Then, smearing his limbs with ashes and wearing a sacrificial thread of hair about his body, the ascetic donned the clothes of the dead man and remained a while in meditation. Thereupon he summoned that most mighty *vetāla* into the corpse by means of powerful spells and proceeded to pay him worship. He gave him, as an offering to a guest, some white human teeth in a skull with fragrant flowers and pastes, and a tribute of human eyes and flesh besides. When he had finished these acts of devotion, he spoke to the king standing at his side:

'Your majesty, may you now make the very deepest of bows, laying your body upon the ground, before this monarch of spells, so that he who grants all wishes may present you with all that you desire.'

When he had heard these words, the king remembered what the *vetāla* had told him and so he replied: 'I do not know how to do this, so please show me first and then I shall follow your worthy example.' As soon as the ascetic fell upon the ground to demonstrate the bow, the king struck off his head with one blow of his sword.

At this a joyous shout arose from the delighted demons who had gathered around, and the *vetāla*, who was pleased, spoke to the king from inside the corpse: 'Your majesty, the lordship over the *vidyādharas*, the fairy attendants of Shiva, which that mendicant sought to gain, shall be yours at the end of your earthly sway. Since I have troubled you so much, choose now a blessing as you desire!'

Upon these words of the *vetāla* the king replied: 'If you are content, what further blessing could be desired? Yet, I shall choose a boon of you so that your words may not be in vain. These varied and pleasant tales which have been told, let them be renowned and honoured throughout the world.'

At the king's request, the *vetāla* pronounced: 'Let it be so; I declare that this garland of tales shall be renowned and honoured amongst men and that they shall lead to welfare. Whosoever shall read but a part of them, or hear it read even, shall be immediately released from their burdens. Nor shall spirits, demons, goblins, witches or *rākshasas* have any power

where these tales are known.' With these words the *vetāla* left
the corpse and returned by his magic power to his own abode.

Thereupon Shiva, surrounded by the gods, appeared, for he
was pleased with the king who bowed low before him, and he
declared: 'A worthy deed this, faithful friend, to kill this false
ascetic who sought to make himself emperor of the *vidyādharas*
by deceitful means. Soon you will become their emperor yourself; meanwhile receive this invincible sword from me, for
through it you will gain everything that you desire.' After pre-
senting the king with the sword, the god Shiva disappeared.
King Trivikrama, for his part, seeing that his task had been
accomplished and the night all but gone, returned and entered
his royal city.